AMERICAN VETERANS

AMERICAN VETERANS

Their Stories of Service and Valor

LT COL JAMES LAWRENCE USAF (RET)

DEEDS PUBLISHING | ATLANTA

Published by Deeds Publishing
Marietta, GA
www.deedspublishing.com

Printed in The United States of America

Library of Congress Cataloging-in-Publications Data is available upon request.

ISBN 978-1-937565-73-2

Books are available in quantity for promotional or premium use. For information, write Deeds Publishing, PO Box 682212, Marietta, GA 30068 or info@deedspublishing.com.

First Edition, 2012

10 9 8 7 6 5 4 3 2 1

DEDICATED TO ALL WHO WORE AN AMERICAN MILITARY UNIFORM

CONTENTS

PROLOGUE

Through the ages, many comments have been made by prominent leaders regarding sacrifices of men and women in uniform. Some of the most often quoted are the words of England's Prime Minister Sir Winston Churchill who always seemed to be able to state things in a rather succinct manner.

When speaking about those British Royal Air Force (RAF) pilots who held off the German Luftwaffe during the Battle of Britain, Prime Minister Churchill made the profound statement, *"Never in the field of human conflict has so much been owed by so many to so few."*

One of those pilots was Sir Douglas Bader who, while outfitted with artificial legs, flew fighter aircraft with the RAF. He had lost both in a crash years before World War II started. Despite his handicap, Bader became a hero and flying Ace by shooting down 22 German aircraft.

Had it not been for those dedicated British pilots such as Bader who put their lives on the line against such great odds, the tide could have turned during that battle, and England could very well have succumbed to the Nazi war machine.

The resulting outcome would have changed the course of world history in ways unimaginable to most of today's Americans.

Many times in the history of American military efforts, the same statement could be made of our own heroes. America as a society owes so much to those *so few* who ensured our survival as a Republic. That could be said not only of veterans of past wars, but ongoing conflicts in which we are involved.

Statistics show the pool of America's so few is becoming ever smaller by the year, relatively speaking. In World War II, some 11.2 percent of our population served in uniform. By the time of the Vietnam War, that figure had dropped to 4.3 percent. At the time of this writing, those fighting our wars were only 0.45 percent of our population.

This means that a lot fewer Americans have "skin in the game," so to speak. Ramifications of this may not be evident for some time in the future, but, it's possible that fewer and fewer Americans may feel a need to sacrifice for the common good, or participate in the defense of the nation. Some may ask, "What can my country do for me, rather than what can I do for my country."

Columnist Cal Thomas once wrote, "Threats to freedom come not just from foreign powers or domestic terrorists. Subtler enemies can enslave us. These come from a focus on *self*—my rights, my pleasures, my money—to the exclusion of what benefits the whole, or as the Founders put it, 'promotes the general welfare.' Who in our increasingly fractured country speaks of the general welfare?" [1] Only time will tell if we stick together, or, because of special interests, become *Balkanized* to our detriment.

This pool of less than one-half-of-one percent of our population serving means multiple deployments and great stresses on those who are actually in uniform today. They and their families truly are paying a heavy price.

Despite those odds, they still volunteer, and they still serve admirably. And, above all, they prevail over our enemies.

The attention of veterans, past and present, to Duty – Honor – Country is amazing, and my intention herein is to capture the essence of the actions of veterans throughout many of the different wars we have faced.

In many cases, these veterans were just ordinary citizens called upon to perform extraordinary service. Many, such as so many of my ancestors, came off farms. Many came from families whose bread winners worked in factories, or in sawmills or gristmills, or some were delivery boys, and many came from families in a wide variety of other circumstances.

In many instances, some became soldiers after getting into some sort of trouble and going in front of a judge who gave them a choice. They could "Join the Army or go to jail," an option often given to troublemakers who opted to join the Army, then became very good soldiers because the discipline and organization of the military were just what they needed.

I have great respect and admiration for all veterans, but I, too, believe that World War II involved "Our Greatest Generation." They defeated the greatest threat to democracy the world had ever seen, and, while doing so, paid a very high price.

World War II was the deadliest conflict in the history of the world. Exact numbers vary, but most agree that more than 416,000 of those in our uniform lost their lives in that conflict (about 9,500 Merchant Marine and 1,900 Coast Guard included) [2]

Not only did the military men and women go off to war in distant areas, but those back home contributed on a massive scale. Many women learned to fly aircraft and served as ferry crews to deliver aircraft from stateside locations to overseas airbases. Others went to work as welders, riveters, or in assembly plants due to a shortage of able-bodied men.

In my hometown of Opp, Alabama, women such as Mrs. Charles Gunnels, Mrs. Howard Teel, Mrs. Joe Benton, Betty Kelly, Bobbie Burgess, Joann Benton, Geraldine Martin, Lucile Pierce, Betty Danley, Mrs. Jane Foreman, Mrs. Ned Moore, Mrs. Quinn Danford and others gathered at the Red Cross Room to make bandages for the wounded. According to the Aug 17, 1944, *Opp News*, they made 759 bandages for war wounds on Monday, August 7, 1944, which was a typical day's work for the group.

Our entire nation had gone on a war footing, beginning by rationing food stuffs and durable goods. Sugar, cooking oils, rubber, gasoline and other petroleum products were rationed so those on the front lines could be given priority for their needs to effectively face and defeat our formidable enemies.

My own mother kept through her lifetime a rationing book issued to her at New Brockton, Alabama, in 1941. She considered it a reminder of how good things could be during years of peace, and how tough things could be in time of war.

It is an interesting artifact for young people of today who are so busy with their iPads, texting, and cell phone calls. They simply cannot imagine such sacrifice, nor can they visualize a time in our future where they might have to give up so many of the amenities they currently enjoy.

After the surprise attack on Pearl Harbor, young men off the farms and out of stores and factories flooded the recruiting stations to sign up for military duty. In my hometown, schoolboy J.T. Benson was driving his 1940 Ford coupe with high school classmates Charles Saunders,

Noah Cowen, Dwight Teel, and Jack Grimes as passengers when they heard a news flash.

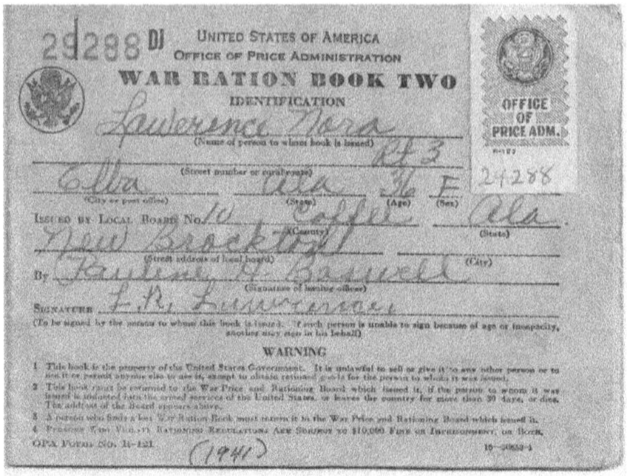

World War II Ration Book (1941), Office of Price Administration issued

As they drove along East Park Avenue about one block east of Fred Lyons' home, the Ford's radio music was interrupted by an announcement about the Japanese attacking Pearl Harbor.

Saunders and the others didn't give it much thought since they were still in high school at the time. Despite their carefree attitude, it wouldn't be much longer before four of the five would be in the military fighting America's greatest war. Two of those four would not survive the war. Their bodies lie in repose on the European Continent to this day, with one in Belgium and one in Normandy.

Immediately after Pearl Harbor, in town after town, double lines of potential soldiers lined up at recruiting stations and backed up for two or three city blocks as they were determined to avenge the damage inflicted by the Japanese at Pearl Harbor. They came from all states of the nation.

Even Hollywood played a big part as those such as Lee Marvin, Charles Bronson, Ernest Borgnine, Clark Gable, Jimmy Stewart, Charles Durning, and many others signed up for duty.

Bronson was a tail gunner on a B-29 flying out of Guam, Tinian and Saipan. Marvin got a Purple Heart at Saipan. Tyrone Power, another star, was a Marine pilot who flew supplies into Iwo Jima, then airlifted out the wounded.

Big band leader Glenn Miller joined the Army in 1942 and had an immediate effect on the morale of troops through his music. Miller's weekly New York City radio broadcasts encouraged volunteers to join the military.

Female movie stars such as Marlene Dietrich, Carole Lombard, and Hedy Lamarr did their part by entertaining the troops and promoting sales of government bonds. Reportedly, Ms. Lamarr even helped with information that contributed to the designs for radio-controlled torpedoes, an amazing innovation for its day.

Ms. Lombard (at the time Clark Gable's wife) died in a plane crash while on a trip promoting the sale of war bonds. She and Gable had created quite a stir in Montgomery, Alabama, when they showed up in 1939 at the Whitley Hotel just prior to the January, 1940 premier of Gable's *Gone With The Wind.* After Lombard's death, Gable enlisted in the Army Air Force and made films to aid in military recruiting. He actually went to gunners school and served as a gunner as part of 8th Air Force, flying five combat missions. Clark Gable eventually achieved the rank of major before resigning his commission after the end of the war.[3]

Baseball players such as Ted Williams, maybe the greatest hitter ever, and "Bullet Bob" Feller, one of the all-time great major league pitchers, joined in the military effort.

Williams became a Marine fighter pilot in WWII, returned to baseball in 1946, then served as a fighter pilot again in the Korean War (a story about his surviving an aircraft crash follows later in the Korean War section herein).

Our country was united in a single tremendous effort to win World War II, with virtually every member of society pitching in and doing their part. For years, the issue was in doubt, but they only redoubled their efforts. They truly are deserving of the title "Our Greatest Generation," and several stories herein will describe their sacrifices.

Praising them does not lessen our respect for veterans of other wars. Other stories herein will describe the service of those in the American Civil War, World War I, the Korean War, the Vietnam War, responses to terrorism, and more modern-day wars. They all contributed their share, regardless of whether the war was popular or lacked solid support from the American public, and they deserve proper recognition.

Some veterans herein grew up in a region near my home. Others were from areas far removed. Regardless of origin, the veterans by-and-large blended into a very effective fighting force and served their country with distinction.

Their stories have motivated me to the point where I feel it is important to record for posterity their extraordinary service, and in some cases, downright heroics. I intend to do that with this treatise.

A special thanks goes to my wife JoAnn and friend Keith Lowery for their assistance with manuscript reviews.

CIVIL WAR ERA

"Duty then is the sublimest word in the English language. You should do your duty in all things. You can never do more, you should never wish to do less."
-General Robert E. Lee, CSA

The War Between the States, or more commonly referred to as the American Civil War, is the bloodiest war ever fought by Americans in uniform. Its gruesomeness can be attested to by a visit to the Antietam National Park where one learns from a Park Guide that, at one small house that had been converted to a makeshift hospital, amputated arms and legs piled up like cord-wood that reached heights equal to the outside eaves of the house.

Casualties in that single day of fighting were 22,000 men. More than 600,000 men and women perished in the conflict that raged between the years 1861 and 1865. Many Americans have ancestors who fought for either the Union or the Confederacy.

Regardless of the beliefs each had in their respective cause, the men and women on each side fought for what they considered at the time a "just cause," to either preserve their way of life, or to render wholesale change to make all persons living in the United States more equal under the law.

Some authors argue that the war was inevitable as *two distinct cultures* were manifesting themselves in progressively diverging political, economic, and social systems.

Several factors led up to the start of the War Between the States. The North was in the beginning of the industrial revolution; the South was still an agrarian society that was dependent upon slave labor, without which a profit could not be made by growing cotton.

The problem of admitting new states to the Union as either slave or free states became a contentious issue and one that could tilt the balance of power in Congress through elections of either pro-slavery or anti-slavery representatives. Other events such as the Dred Scott decision and John Brown's raid on Harper's Ferry stirred up things.

Then, the election of Abraham Lincoln in 1860 was for many Southerners the proverbial "straw that broke the camel's back," further stoking the fires for secession. Though Lincoln did not actively identify himself with abolitionists, the abolitionists identified with and supported Lincoln's bid for the highest office.

Finally, States' rights was a really big issue and became the rallying cry for those in the South who favored a split. The South feared a strong, centralized government such as the North offered that might threaten their commerce and way of life. [1]

"Between 1789 and 1861, a new people were unconsciously created by changes in environment. It was the unconsciousness of this fact that led to the war. The economically-minded North felt that prosperity depended upon maintenance of the Union; the agriculturally-minded South felt otherwise; not that Southerners did not realize the value of the Union, but they scorned to place prosperity above individual freedom. It was the South that had led the rebellion for freedom, and to the aristocracy of the South freedom was beyond all price." [2]

Those in the North expected a quick victory. So did the South, although an objective look at war-making capacity gave the decided edge to the Union, and that positive outlook was reflected in the populace.

"The North had better reason for optimism. It had more than a two-to-one edge in manpower and almost all of the nation's industrial plants. There were no railroad repair shops in the South, no mills that produced woolen cloth, no shoe or glass factories, and no facilities for producing needles, pins, machetes or even salt. The Confederacy, however, was the world's leading producer of cotton and counted on English and French intervention to prevent the shutdown of European mills. The Confederates also were fighting for the most part on their own territory." [3] When the war was over, the South was ravaged and recovery would be a long time coming. Some 258,000 [4] Confederates had died in the fighting (many were from disease), a figure that represented close to 18 percent of Southern males between ages 13-43.General Sheridan in October 1864 bragged to General Grant, "I have destroyed over 2,000 barns filled with wheat, hay and farming implements; over 20 mills filled with flour and wheat; have driven in front of the Army over 4,000 head of stock and have killed and issued to the troops not less than 3,000 sheep...all the houses within an area of five miles were burned."

The North suffered more than 360,000 deaths [5] which represented about six percent losses in the same age groups. Some 50,000 civilians in the South lost their lives, and more than 200,000 (both blacks and whites) were left homeless.

MATHEW M. "MACK" RUSHING

"So severe were his wounds that a good portion of his left arm was amputated"

Mathew M. "Mack" Rushing was this author's maternal great grandfather, born circa 1829, the first child of William Cain and Mary Rushing, at Robertson Cross Roads, Montgomery County, Alabama.

His father, William Cain Rushing, was prominent in the area, twice serving as a Pike County Justice of the Peace on two commissions (1847 and 1850).

When the American Civil War began, Mack Rushing enlisted at Elba, Alabama in the Confederate Army, Company A, 33rd Alabama Infantry, in February, 1862. He fought bravely for the South, but was severely wounded by a musket ball at the Battle of Perryville, Kentucky.

The bloody battle occurred on October 8, 1862, with Confederate General Braxton Bragg commanding Southern forces, and General D. Buell directing Union forces. The battle came after Union General Grant had won a narrow victory over Confederate forces at Shiloh and had then turned his attentions toward Vicksburg.

General Buell of the Union had headed toward Chattanooga, but turned to follow Bragg who had marched toward Kentucky, hence the fight at Perryville.

In the battle, Union forces suffered 4,276 casualties (894 killed, 2,911 wounded and 471 captured or missing in action). CSA forces suffered 3,401 casualties (532 killed, 2,641 wounded, and 228 captured or missing in action). [1]

So severe were Mack Rushing's wounds that a good portion of his left arm was amputated. He was discharged at Chickamauga, Tennessee, after his surgery and returned to the Zoar Church area, roughly eight miles northwest of Elba, Alabama to farm for a living.

After the long trek home to the Elba area, a friend rigged up a special harness so Mack could plow his mules and oxen with one hand, which was quite a feat.

He was also a crack shot with a rifle, using only one good arm. The families of his day and age depended a lot on wild game, and it was a good thing he could still do some hunting to help feed such a large family (of 11 children).

Mack's wife was Sarah Jane Amerson whom he married in 1853. Both are buried in the Zoar Church Cemetery. One of their sons, Joseph Thomas Rushing, is this author's maternal grandfather.

Several of Mack's brothers also served in the Confederate Army. John Eager Howard Rushing was a Private in the 53rd Alabama Regiment, Company B (at Troy). He was named for John Eager Howard who had been a colonel in the Continental Army, elector to the Continental Congress of 1788, Governor of Maryland, a state senator, and later a United States senator.

Joseph T. Rushing entered service at Fort Mitchell, Alabama on July 13, 1861, as part of the 15th Alabama Regiment of Alabama Volunteers. He was wounded at Cold Harbor and on the Darbytown Road (he was Orderly to Oates).

At the battle of Cold Harbor (east of Richmond), the South's losses were slight. The North suffered 1,100 killed and 4,517 wounded. The battle did not turn out well for General Grant. Grant's plan was for a frontal assault on General Lee's forces, and, after it was postponed 24 hours, began at 4:30 a.m. on June 3, 1864.

The delay was a mistake that gave Lee time to shore up his defenses. Another mistake on the part of Grant was to order an attack all along the line in place of massing his guns opposite a fraction of Lee's front, and then after a heavy bombardment, assaulting this fraction.

Each of Grant's divisions was taken in enfilade, as well as decimated by frontal fire, and all were so severely handled that it was decided in less than an hour. [2]

Joseph T. Rushing was one of those counted among the *slight* losses/wounded of Lee's forces at Cold Harbor.

Thomas A. Rushing was in the 60th Alabama Infantry Regiment, Company D. He was killed in the trenches on August 11, 1864.

Solomon James Rushing enlisted in Montgomery 27 June, 1862 in Company C, 60th Legion. He was wounded at the Battle of Chickamauga and was sent home to recover. He returned to battle and was again wounded, then spent the rest of the war in the Confederate Soldiers of America General Hospital in Farmsville, Virginia. [3]

Mathew "Mack" Rushing and his brothers were typical of many young men of the South—poor farm boys who were patriotic to their area of the country and to a cause they believed in. Whether right or wrong, they fought with all their might to prevail. Just more examples of *ordinary citizens performing extraordinary service.* That's the American way!

WILLIAM JOSIAH "DIAMOND JOE" WARD

"Twenty five grand would pay a lot of 'Johnny Rebs' for a long time"

William Joseph "Diamond Joe" Ward (and wife Mary Purvis Ward)

When General Robert E. Lee surrendered to Union General Ulysses S. Grant at Appomattox, the Civil War came to a close, or it should have. Yet, some in the Confederate Army had other ideas, which is where the story of Diamond Joe Ward comes into play.

William Josiah "Diamond Joe" Ward was born to James Benjamin and Elizabeth (English) Ward on March 23, 1830, in Dale County, Alabama. He was called Diamond Joe because he always marked a diamond on each of his possessions.

Diamond Joe was a very intelligent man who believed strongly in education. It was well known in family circles that he could recite most of the Bible by memory. At one time in his life he built a log school house on Seven Run Creek near Bruce, Florida.

On January 1, 1850, Joe married Mary Apseybeth Purvis (who lived until November 15, 1913). They had nine children—four sons and five daughters—with many of those descendants still living in the Florida Panhandle.

When the Civil War broke out, he proudly served in the 53rd Alabama Infantry Regiment.

"The 53rd Alabama Cavalry Regiment, Partisan Rangers, was organized by increasing the 1st Cavalry Battalion to regimental size at Montgomery on 5 November 1862. Recruits were from Autauga, Coffee, Coosa, Dale, Dallas, Lauderdale, Lowndes, Macon, Monroe, Montgomery, Pike, Tallapoosa, and Wilcox counties. It proceeded in a few weeks to Mississippi. In moving from Columbus to Decatur, in Lawrence, a portion of the regiment was there equipped and proceeded to join General Earl Van Dorn.

This battalion was in the fighting at Thompson's Station, and at Brentwood. The regiment was engaged in the fight with Union General Grenville Dodge at Town Creek and in the pursuit of Union Col. Abel Streight. Soon after, the 53rd joined the main army at Dalton as part of General Moses W. Hannon's Brigade, General John Kelly's Division. It operated on the right of the army as it fell back towards Atlanta and was engaged in constant duty.

When Union General William T. Sherman reached Atlanta, the 53rd was the principal force engaged in the daring raid in his rear, whereby a valuable train was destroyed. It was then at the heels of Sherman as he devastated Georgia and the Carolinas, and it took part in the last operations of the war in that quarter. It surrendered a small number with General Joseph E. Johnston at Durham Station, Orange County, NC, on 26 April 1865." [1]

Diamond Joe Ward had done his duty, but, after General Robert E. Lee surrendered, Ward was ready to return to a peaceful existence and get on with a more productive life.

However, other Confederates such as a group of soldiers from a Mississippi regiment had other ideas until they encountered Diamond Joe (His story was told to this author by his great-grandson Ralph Odell Ward who has a story of his own military service at a later point in this accounting).

The story goes that at war's end, within the Confederate Treasury at Richmond, Virginia, there was a hidden stash of over $1,000,000.00 in gold coin. The Confederate holdouts from Mississippi raided the Treasury and grabbed enough gold to continue a resistance back home in Mississippi. The amount taken was around $25,000.00, a goodly sum back in 1865. Twenty five grand would pay a lot of "Johnny Rebs" for a long, long time!

Later word was they buried some of the gold in either North Carolina or South Carolina, but nothing was ever found. In actuality the Rebels made it as far south as Gainesville, Florida with the looted treasure.

Diamond Joe Ward got wind of their movements and hatched a plan to derail their attempt to continue the war. He had had enough of war during his time in the 53d Alabama Infantry, he wanted all the killing stopped, and he felt it was foolish to continue to resist. Ward wanted to push efforts toward higher education for those in the South.

Just below Ebro, Florida, at a place called *Cow Ford* on the Choctawhatchee River, Ward in a fit of sabotage rigged up the ferry carrying the gold, soldiers, and horses so that it would sink when it sailed further downriver. When it sank, there was a large loss of many horses aboard the ferry, not to mention all those gold coins that went to the muddy bottom.

The Mississippi Confederates made it out alive, but had no way to continue home, not to mention that all their gold was at the bottom of the river, along with a significant number of drowned horses. They were very dispirited.

The next day, Ward got the men on wagons and took them over to Freeport, Florida, where he was then able to get them on a boat to take them on to Mississippi. They were none the wiser for Joe's perfidious deed.

A talent that Diamond Joe Ward had was that he was able to hold his breath for at least three minutes. Over a period of weeks, he dove repeatedly on the ferry wreckage until he was able to recover most of the gold.

He then used the recovered Confederate Treasure to support a school of higher education in DeFuniak Springs, Florida. Many of his children and grandchildren were able to pursue their education at the school.

Over the decades, the family swore all who heard the tale to secrecy. Prior photo of Ward and wife came from a genealogy website referenced:[2]

Ralph Ward's revealing the tale to this author is the first time it has been told to someone with the intent that it would become public knowledge.[3]

WILLIAM MIDDLETON E. RICHBURG

"I still have faith in our all getting back home in a short time."

Private William Middleton E. Richburg, born 1835 in Pike County, Alabama, was a Confederate States of America soldier fighting for his beloved Southern States as part of Volunteer Company C of the 37th Alabama Regiment. Captain T.C. Kendrick was initially his Company CO. He enlisted on May 5th and was mustered in at Auburn, Alabama on 13 May, 1862.

Not a lot is known about his service, other than through a deciphering of letters sent home to his wife, Mary, during his time in war. There is however one major bit of known information. He never returned home alive, nor did his body come back.

Some of those letters give insight to what was going on in the fields of encampment after the letters were read, translated and typed for Richburg's descendants who became interested during a genealogical search by the family.

We know that William married Mary Harrell, daughter of James and Matilda (Shipp) Harrell, and they had one son, Jack Manning Richburg, and one daughter, Willie Elizabeth Harrell (Richburg) Saunders, who then had a son—Lawrence Leonard Saunders (whose friends affectionately called him 'Double L"). Through Double L they had a great grandson, Charles Saunders, who served in World War II. Charles' story comes later.

This author interviewed Charles Saunders in Troy, Alabama to capture and piece together the story of Richburg's role in the Civil War. He was kind enough to provide copies of several of William Richburg's Civil War letters. Charles' sister-in-law, Dot Saunders, was also very helpful in tracking Mr. Richburg's story.

The letter of June 8, 1862 indicated to his wife, Mary, that the Confederates were near Columbus, Mississippi, with four regiments there, and some 90 miles from Corinth, site of a later large battle. He had seen one man (named Gilmore) drown at Montgomery (likely in the Alabama River). They spent two and one half days on a boat.

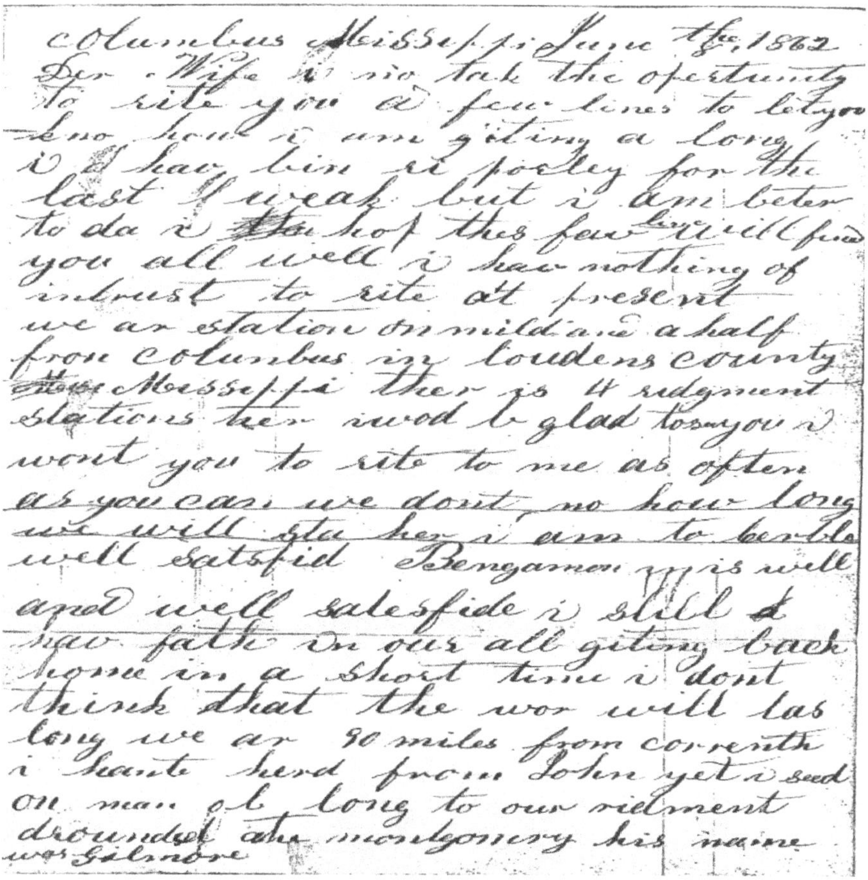
columbus Missisippi June the 8, 1862
Der Wife i no tak the opertunity
to rite you a few lines to let you
kno how i am giting a long
i hav bin ri pocley for the
last week but i am beter
to da i hop thes few lines will find
you all well i hav nothing of
intrust to rite at present
we ar station on mild and a half
from columbus in loudens county
Missippi ther is 4 ridgment
stations her i wod b glad to see you i
wont you to rite to me as often
as you can we dont no how long
we will sta her i am to be to
well satsfid Bengamon is well
and well satesfide i still
hav faith in our all giting back
home in a short time i dont
think that the wor will las
long we ar 90 miles from corenth
i haint herd from John yet i seed
on man b long to our rielment
drounded at montgomery his name
was Gilmore

Richburg's handwritten letter June 8, 1862, Columbus, Mississippi, Pg 1
Columbus, Mississippi June the 8th, 1862

Dear wife, I now take the opportunity to write you a few lines to let you know how I am getting along. I have been right poorly for the last week, but I am better today. I hope these few lines will find you all well. I have nothing of interest to write at present. We are stationed a mile and one/half from Columbus in Lowndes County, Mississippi. There are four regiments stationed here. I would be glad to see you. I want you to write to me as often as you can. We don't know how long we will stay here. I am too crippled to be well satisfied. Benjamin is well and well satisfied. I still have faith in our all getting back home in a short time. I don't think that the war will last long. We are 90 miles from Corinth. I haven't heard from John yet. I saw one man who belonged to our regiment drown at Montgomery; his name was Gilmore. We were on the boat at Montgomery two days and one half. Direct your letters thus to Wm M. Richburg in the care of Captain Kendrick 17th Alabama Volunteers. I fell on the 19th and am still a mending. Today for the first time I am able to wash my clothes. I request all my friends to pray for me. Nothing more at present. Only I remain your husband till death.

From William M. Richburg to Mary Richburg.

Richburg's next letter of July 7, 1862, from Columbus, Mississippi, indicated he had not been doing well, but was mending. There was a good bit of sickness among the soldiers in his regiment, there had been another fight at Richmond, and he expected to be home in September.

(Note: Author found a common theme in reading many Civil War letters written by soldiers in the field. They *almost all* included a prediction of being home in the near future, a prediction that in most cases was grounded more in hope than in reality).

Richburg also admonished his beloved Mary to write to him more frequently, complaining that he had only received one letter from her since he left home.

Columbus Mississippi July 7, 1862

Dear wife, I take the pleasure of writing you a few lines to let you know how I am getting along. I have been poorly though I am mending soon. I hope these lines find you all well. I have nothing of interest to write at present. I have got no money yet though I don't know when I shall get it. I have no news to write. There is a good deal of sickness in our regiment, though we have not lost nary one out of our company. Benjamin has been sick with a cold. He hankered down sick. He is mending. I want you to write to me. I (long?) to hear of my friends. We have had another fight at Richmond (Rightmon). I think that will about wind up the show. I think we will all be home by the last of September. So I will close. Nothing more at present. Only I remain your husband til death.

From William M. Richburg to Mary Richburg
To: Mrs. Mary Richburg
At Home
Pike County, AL

Direct your letters thus to Wm M. Richburg, in the care of Capt T. C. Kendrick, Co. C, 37 ALA Volunteers
(Deciphered Richburg letter of 7 July 1862)

The following letter sent by Private Richburg (which was started August 15, 1862, then completed on August 22) gives an accounting of several in his Company C that are sick, or are recovering from recent sickness. He praises his leaders as being "noble" (especially a Lieutenant James Burnett) and "Christian-hearted" men such as a "good old colonel," and once again implores his wife to write as often as she can, *once a week if possible* (another common theme seen throughout most all letters written by Civil War soldiers).

One man is sick with Typhoid Fever, and Richburg supposes his company of troops is near the enemy in large forces, and his officers say they are liable to be attacked everyday.

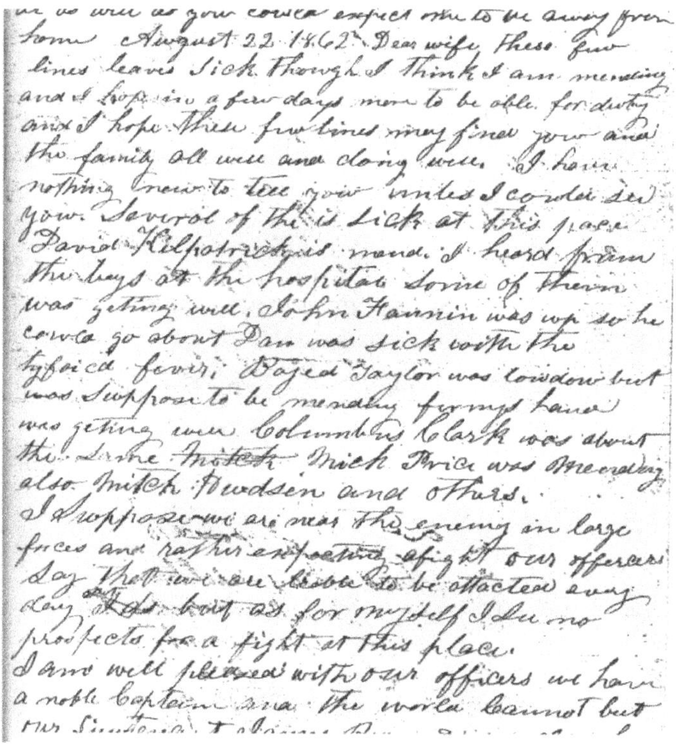

[illegible] as well as you could expect me to be away from
home August 22 1862 Dear wife these few
lines leaves sick though I think I am mending
and I hope in a few days more to be able for duty
and I hope these few lines may find you and
the family all well and doing well. I have
nothing new to tell you unles I could see
you. Severol of the is sick at this place
David Kilpatrick is mend. I heard from
the boys at the hospital some of them
was geting well. John Hannin was up so he
could go about Dan was sick with the
typhoid fevir; [illegible] Taylor was lowdow but
was suppose to be mending for my hand
was geting well Columbus Clark was about
the same ~~Mitch~~ Mich Price was mending
also Mitch [illegible] and others.
I suppose we are near the enemy in large
forces and rather expecting a fight our officers
say that we are liable to be attacted every
day [illegible] but as for myself I see no
prospects for a fight at this place.
I am well pleased with our officers we have
a noble Captain and the world cannot beat
our [illegible]

Will Richburg's original Saltillo, Mississippi letter of 15 Aug/22 Aug 1862

Satillo, Mississippi August 15th 1862

Mrs. Mary Richburg. Dear Mary, I now seat myself to drop you a few lines by which you may know that I have not forgotten you. I am very well satisfied as well as you could expect me to be away from home. August 22 1862 Dear wife, These few lines leave me sick though I think I am mending and I hope in a few days I will be able for duty and I hope these few lines may find you and the family all well and doing well. I have nothing new to tell you unless I could see you. Several of them are sick at this place. David Kilpatrick is mending. I heard from the boys at the hospital, some of them was getting well. John Fannin was up so he could go about. Dan (Dave?) was sick with Typhoid Fever. Boyd Taylor was lowdown but was supposed to be mending. For myself was getting over Columbus. Clark was about the same. Mich Price was mending also Mitch Hudson and others. I suppose we are near the enemy in large forces and rather expecting a fight. Our officers say that we are liable to be attacked every day, but for myself I see no prospects for a fight at this place. I am well pleased with our officers. We have a noble Captain, and the world cannot beat our Lt. James Burnett. We also have a good ole Colonel, he is a good Christian hearted man. I want you to write me soon and give me all the news. Let me know how everything is getting along, and tell me what they say about the _____? crop. I suppose they are taking the war_____? I want you to write to me how your crops look and whether you think you have enough to do you or not. You must excuse me for not writing oftener for my paper is scarce and hard to get. But I want you to write to me as often as you can, once a week if possible. Well I will come to a close. Give my respects to all injured after me if there be any and expect the same.

From Your Husband
Will Richburg
Satillo MS one10
Hwy 24
Mrs Mary Richburg
New Providence
Pike County
Alabama

Pvt. W M Richburg
37 Reg Ala Vol
Company C

(Richburg's letter of 15/22 Aug 1862, as deciphered)

In the following Richburg letter of September 4, 1862, he writes that he is better and that his company is heading up to Corinth, Mississippi (possibly as soon as the following day) [Note: Corinth was a major battle in the Civil War].

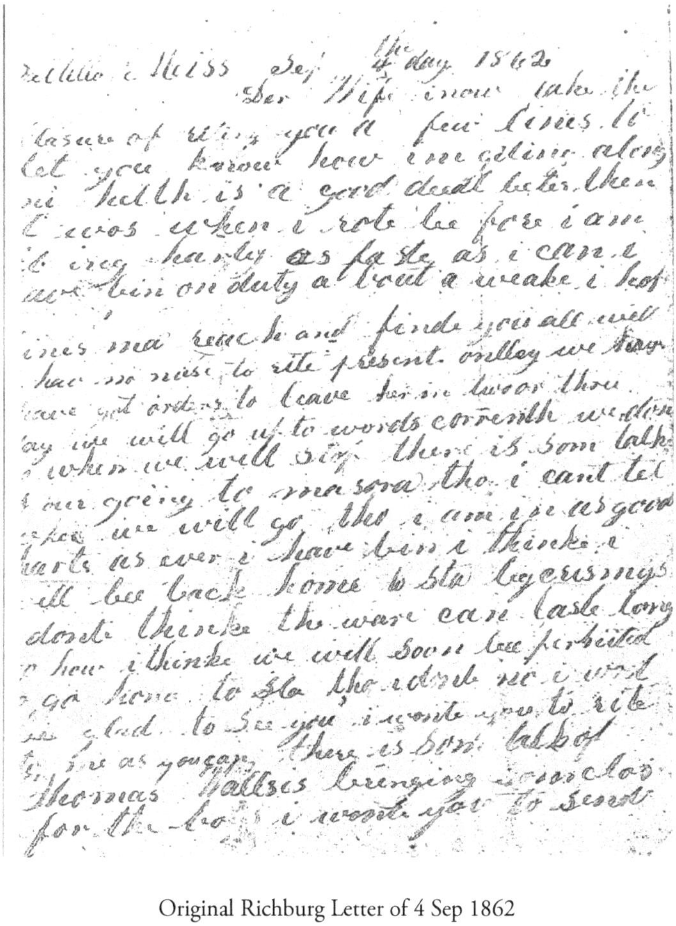

Original Richburg Letter of 4 Sep 1862

Satillo, Mississippi September 4, 1862

Dear wife, I now take the pleasure of writing you a few lines to let you know how I am getting along. My health is a good deal better than it was when I wrote before. I am getting hardy as fast as I can. I have been on duty about a week. I hope these lines will reach and find you all well. I have no news to write, only we have orders to leave here in two or three days. We will go up towards Corinth. We do not know when we will stop.

There is some talk of our going tomorrow though I can't tell when will go though. I am in as good health as ever I have been. I think I will be back home to stay by Christmas. I don't think the war can last long no how. I think we will soon be permitted to go home to stay, though I don't know. I would be glad to see you.

I want you to write to me as soon as you can. There is some talk of Thomas Wallis bringing some clothes for the boys. I want you to send one set of clothes. I need a coat and hat. I need socks and gloves for winter.

I sent you ten dollars by John Macktes, and I will send you money as soon as I (get more?). It is now night and I must close; so, nothing (more?) at present only *I remain your husband till death.*

William M. Richburg.
(Letter of Sept. 4, 1862, deciphered)

He predicts the war will be over soon, and that he will be home by Christmas! He asks Mary to send a set of clothes, including a coat and a hat. Also socks and gloves for winter. He tells her he sent her $10 by John Macktes and will send more as soon as he gets more.

Richburg closes by adding "I remain your husband until death."

This last comment on most of his letters was a prophetic closing statement. A search of National Archives records by Richburg's relatives revealed that he was suffering from pneumonia and was killed in 1862 in Corinth, Mississippi. Dot Saunders later found other information that revealed William Richburg was buried near Corinth with other CSA soldiers who were killed during that battle.

More than a century later, when a four-lane highway was constructed, it was built on top of some of the graves, many of which were unmarked.

According to her, after so many were killed on the battlefield, predators of the Civil War era would search the bodies before burial, remove any identification, and then use the documents to fraudulently apply for veterans benefits for the family, but pocket the money themselves.

Richburg's family was in dire straits after the war, but never received any benefit from a soldier's pension (per Ms. Dot Saunders). [1]

The Alabama Archives and History records (accessed 26 Feb 2013) show Mary Richburg filed a claim for a one-time widow's payment with the Confederacy through the offices of J.G. Hedgeman, in Richmond, Virginia. The claim was filed May 18, 1863. There is no other information as to whether or not the claim was honored.

JONATHAN P. & JOHN THOMAS HENDERSON LOW(E)

"I pray for you nearly every breath I draw"

John Thomas Henderson Low(e)—family photo

Jonathan P. Low(e) of Lowndes County, Alabama, was a Confederate States of America soldier, a private in rank, with the 3rd Regiment, Alabama, company unknown.

He and wife Lucy sacrificed greatly to the Confederate cause. Not only did Jonathan P. Low(e) serve, but four of their sons also served. One son that we have information on was John Thomas Henderson Low(e).

It's known that John Thomas Henderson Lowe survived the war and returned home. His father Jonathan P. did not. This letter was one from Lucy to Jonathan, 1 June, 1862.

From the book *Heroines of Dixie: Confederate Women Tell Their Story Of The War*, was the letter from Lucy Lowe to her husband.[1]

Stories passed down of family history indicate that he died of non-hostile causes, thought to be the result of contracting Smallpox, a not uncommon occurrence among Civil War soldiers.

After a long and exhausting march, he was found dead the next morning leaning up against a tree. Rigor mortis had already set in, and his fight was over. The fight went on for the four sons, including for John Thomas H. Low(e).

In Jonathan's absence, plus the absence of their four older sons, she had managed the crops, fed and clothed the small children, and prayed for her husband with nearly every breath she drew (see letter). She also discussed problems one of their sons, George, was having from having been in war. Back then, people called it "shelled shocked." These days, we refer to it as PTSD.

Lucy and Jonathan's son, John Thomas Henderson Low(e), was born on January 20, 1843. Confederate military records showed a postal address for him in Honoraville, Alabama.

He was doing well until the Battle of Franklin, Tennessee. There, on November 30, 1864, he was shot in the shoulder and captured by Union forces. He was processed as a POW under forces of Major General Thomas, Dept. of the Cumberland, then transferred to Captain S.E. Jones A.D.C. Louisville, Ky. on December 2nd.

3. LUCY LOWE—"I WANT TO SEE YOU SO BAD"

Lucy Lowe, wife of John Lowe of Lowndes County, Alabama, gave four sons and a husband to the Confederacy. During their absence she managed the crops, fed and clothed the small children at home and prayed for her husband "nearly every breath she drew."

lounds County June 1 1862

Dier husban I know take my pen in hand drop you a few lines to let you no that we are all well as common and I am in about the same helth that I was when you left I hopes these lines may find you the same I receved a leter from you the 29 of May and you sed that you had not got many leters from me yet I have sent you fore leters besides this one George has got home he got hear the end of May he is very lo yet the boys has come home to sea me on a furlow and stade 10 days they started back yestrday to the camp they dont no whare they wil git to from their they are station at arbon [Auburn] above Mongary [Montgomery] John my corn is out know and I have not drawed any thing yet but I hop I wil my crop is nice but pane hes quit and left my crop in bad fix but the neighbors ses they will help us you sed you wanted me to pray for you as for prayers I pray for you all of the time I pray for you nearly every breth I draw and I want you to pray for your self I have give the bys 13 dollars and I bought some 9 bushels of corn and that is all that I give him for his work and I have got 20 dolars yet George is very bad yet and he dont no whether he wil ever get able to go back the sargent give him furlo to stay at home til he was able to rejoin his company and he ses for you to try to get him a discharge from the head one for he never will be able to go back agane and I want you to get one if you can George ses that you can get one any time you want to Sister ses she wants to sea you and

150 THEY CALLED THEM "GREAT DAYS"

kis you Your baby is the pertyest thing you ever saw in your life She can walk by her self and your little gran son is perty as a pink and growes the fastest in the world you must come home and sea all of your babyes and kis them I have got the ry cut hook sent and cut it for nothing Your old mare is gone blind in one eye and something is the matter with one of her fet so she cant hardly walk your hogs and cows is coming on very wel I want you to come home for I want to sea you so bad I dont no what to do I must come to a close by saying I remane your loveing wif until deth You must write to me as soon as you get this leter goodby to you

Lucy lowe to John P. lowe

Letter from Lucy Lowe to husband Jonathan P. Lowe

Low(e) was moved north and ultimately held as a prisoner of the Union at Camp Douglas near Chicago which was open as a prison camp from 1862-1865.

After being freed, John Thomas H. Low(e) told the story of how he stayed alive when some 6,000 other Confederate prisoners perished at the camp.

He was a good cook, and he made sure he cooked well enough that he was always the one chosen to cook for the Union Army officers. As long as the food was good, they had a reason to keep him alive, not a bad strategy for one in dire straits.

J.T.H. Low(e)'s second wife was Harriet Elizabeth (Vickery) who was born December 29, 1839.[2]

WORLD WAR I

"Woodrow Wilson had tried hard to keep America out of World War I. Not only was he reluctant to send his countrymen to die in a war whose moral basis was unclear, but peace was better business. While the European powers destroyed one another, the United States grew rich off exports and international banking. But by early 1917, after making repeated efforts to bring the warring countries to the bargaining table, Wilson felt he had no other choice.

"Two factors forced his hand. The first was Germany's war on commercial shipping... the Germans had attacked unarmed ships in British waters...by mid-March, German submarines had torpedoed three U.S. nonbelligerent vessels, as well as some 1.3 million tons of Allied ships. Hundreds of Americans were among the casualties.

"Then U.S. agents intercepted the so-called 'Zimmerman telegram'—a secret message (in all likelihood a setup) from the German foreign minister ordering his ambassador in Mexico to offer Mexico aid in retaking New Mexico, Texas, and Arizona from the United States.

"In April, proclaiming that 'the world must be made safe for democracy,' Wilson asked Congress to declare war." [1]

One dramatic event, the sinking of the *Lusitania* (May 7, 1915), a British *Cunard* Company liner, had claimed 1,195 lives, including 123 Americans. The hue and cry from this had created great emotionalism encouraging America's later entry into the war.

The *catalyst* for the war starting was the assassination of Archduke Ferdinand of Austria on June 28, 1914, in Sarajevo, by a Serbian nationalist student. The *reason* the world came to be involved so quickly can be traced back to the numerous interlocking treaties and alliances among the participant countries, with the result that the largest war in history was soon raging on three continents.

The industrial revolution would greatly affect World War I as machinery and products would play a big part in the effectiveness of the warring sides.

On August 6, 1914, when Germany began moving troops to the Western Front, some 550 trains per day crossed the Rhine River bridges with troops; some 3,120,000 soldiers crossed in eleven thousand trains. [2]

WWI also marked the beginning of the use of huge battleships. "Warship design and armament were transformed; the '*Dreadnought*' ushered in the new era of the all-big-gun battleship." [3]

Tank warfare came into use, and horrible chemicals manufactured on a large scale would come to be used against those in uniform.

The human body is not equipped to ward off the effects of such chemicals, so the result was a very ugly outcome for many young men on both sides of the conflict.

Many had never heard of such chemicals, nor had they seen their effects until they encountered their debilitating effects firsthand, either in a foxhole or hunkered down with the rats in some long, muddy trench near the frontlines.

The airplane, as basic as it was, came to be used in observation, intelligence gathering and pursuit roles. The widespread use of trains and motorized trucks to move large numbers of troops ensured even larger and more deadly battles. Casualties piled up in huge numbers. Gunboats attacked each other on the high seas, and Britain's superior Navy soon blockaded Germany. In return, Germany turned to the use of *U-boats* that quickly began sinking both military and commercial vessels, including actions that killed Americans. The battle was on, and it proved deadly and expensive to all involved.

When it was finally over, the *Treaty of Versailles* that ended the war was so one-sided that Germany could not pay the outrageous reparations and conditions demanded by France and England, such as giving up half of its iron and coal, forfeiting 25,000 square miles of territory and $40 billion currency, and yet be excluded for consideration for the new League of Nations so advocated by American President Woodrow Wilson.

Animosities sprang up in a terrible post-war German economy, which then sowed the seeds that allowed Adolf Hitler to come to power and precipitate the beginning of World War II in a mere two decades.

My own father had told me stories that he had read about Germans rolling a wheelbarrow of German *Marks* to their local grocery store just to buy a single loaf of bread. It's hard to imagine that kind of inflation, but it happened. Most societies seek relief from that type condition, and Germany's answer was to turn to Adolf Hitler and his proposed "final solution."

The Allies had won a military victory on the WWI battlefield, but they *lost the peace* thereafter because of their desire (especially France and England) to punish Germany and make sure she could never build a military machine capable of waging war again.

In Flanders Fields

By: Lieutenant Colonel John McCrae, MD (1872-1918)
Canadian Army
(Poem Composed in the Battle of Ypres, Spring 1915)

In Flanders Fields the poppies blow
Between the crosses row on row,
That mark our place; and in the sky
The larks, still bravely singing, fly
Scarce heard amid the guns below.

We are the Dead. Short days ago
We lived, felt dawn, saw sunset glow,
Loved and were loved, and now we lie
In Flanders fields.

Take up our quarrel with the foe:
To you from failing hands we throw
The torch; be yours to hold it high.
If ye break faith with us who die
We shall not sleep, though poppies grow
In Flanders fields.

WILLIE LEE RUSHING

"A dollar a day to expose oneself to poison gas, bullets & bayonets"

Will Rushing, lower right (with cane), with brothers Dan and Frank

A maternal uncle of this author, Willie Lee Rushing was born August 16, 1898, to Joseph Thomas and Hettie (Floyd) Rushing in the Zoar Church community some eight miles northwest of Elba, Alabama. He was a grandson of Mathew "Mack" Rushing.

Times were tough back in those days, and military service was a way to make a little money. He told family he enlisted at a very early age, then re-enlisted when he was 19 years old.

He entered the Army at Fort McPherson, Georgia, then was sent for training in the Artillery field to Camp Travis, Texas. During the war, he recalled that he was assigned to the 5th Field Artillery in France. The Alabama Archives and History records (accessed on 26 Feb 2013) show he served overseas May 27, 1917 to July 21, 1919, and indicates he at one time was with Battery F, 19th Field Artillery.

It was in this First World War that trench warfare and tanks were introduced on a large scale. Chemical warfare was also used extensively, with mustard gas, chlorine, tear gas, and phosgene being employed on thousands of unfortunate troops in the trenches. Over 2,500 Alabamians died in WWI.[1]

Will's son, Don, remembered his dad saying that he spent a lot of time in foxholes during his time in combat in France. Fortunately, he was never gassed by the enemy, but heard about others who were.

As an Army private, Rushing was paid one dollar per day. Think about that—a dollar a day to expose oneself to poison gas, bullets, and bayonets—incredible! A private first class got three dollars more per month. With another promotion, a corporal received the "unbelievable" sum of $36.50 per month.

When he was discharged from service, he was ill when he headed back stateside. [2] He came home on a cot with a blanket, and was admitted first to the Pensacola VA Hospital for two weeks, then was sent to the Montgomery VA Hospital. He ultimately spent nine months in the hospitals.

Using war bonus money afterward, he bought land for only $3 per acre in Red Bay, Florida, south of DeFuniak Springs.

He later worked in the timber business, then operated a country store along Highway 81 in Red Bay for some 30 years. He was a member of the Masons and was dedicated to his Presbyterian Church. He died January 5, 1991.

When Uncle Willie visited his nephews in Alabama, he always brought along from his Red Bay store packages of what he termed "shipyard cookies."

These were marshmallows on cookies with coconut sprinkled on top (real name: Sno-Balls), but he termed them shipyard cookies because so many welders, pipefitters, and assembly workers en-route to the Panama City shipyards during World War II stopped and bought them as desserts for their brown bag lunches.

We were delighted to get something so delicious.

When he was no longer able to run the general store, he passed it along to his daughter, Murlene Gomillion.

Uncle Will's store eventually changed hands several times after his ownership, and it closed down for a period of time in 2009. At one time it almost burned completely, putting those in the vicinity of Red Bay in the position of having to drive 25 miles just to get a loaf of bread.

But, now, thanks to the efforts of a group in the area that bought the business, it has been resurrected as a restaurant and enjoys a clientele from far and near.

MARION WASHINGTON LOWERY

"He mostly picked up dead bodies off the battlefield"

Private Marion Washington Lowery WWI Uniform (family photo)

Private Marion Washington Lowery was born near Georgiana in Butler County, Alabama on August 2, 1890, to George and Lou Anna (Pitts) Lowery.

As a young man, he grew up on a farm and did more than his share of manual labor, working in the cotton and corn fields of his parents, and occasionally working for neighbors who needed a helping hand.

As World War I raged in Europe, Lowery on July 9, 1918, was drafted into the Army, inducted at Greenville, then processed and entrained for duty at Camp Shelby, Mississippi on order #1535. After basic training, his advanced training was as a medic. [1]

A card he sent back home read, "My dear father and mother. I am all o.k. at the present, and hope these lines will find you all the same. You can write to me at the following address: Marion Lowery, 113 Sanitary Train, 152 Ambulance, New York AEF (AEF = Allied Expeditionary Force). Please hand this to mama for me."

Lowery was trained as a medic and initially forward deployed to England. Records show he deployed for overseas duty on October 6, 1918 and was there until February 16, 1919. From England he went on into the battle areas of continental Europe, evidently France, from the looks of the postcard.

Those members of his family who recalled any conversations with him about his war time duty indicated he (in his words) "...mostly just picked up dead bodies" off the battlefields. Mortality rates were high. Medicine was so basic in 1918-1919 that any soldier with more than a superficial wound was subject to dying from infection, if not from bleeding out from the projectile itself.

While in England, Marion met a pretty young nurse who after the war sent multiple photos. The long distance romance never panned out. The distances were just too great. Marion Washington Lowery married Lillian Mae Lowe after returning to Alabama.

Lowery's ship arrived back in the United States on February 16, 1919. He was issued an honorable discharge on February 28, 1919, having done his duty for his fellow countrymen. [2]

After the war, Marion Lowery picked up right where he left off, resuming his life as a farmer, getting married, and raising children.

One of those children was Sergeant Lee Jack Lowery who would serve in the Marines in World War II, with duties as an Armorer-Turret Gunner aboard a PBJ-1H, a Navy version of the B-25 Bomber. His service would be in the Marshall Islands (Kwajalein) with Marine Bombing Squadron 613. Sergeant Jack Lowery's story comes later.

Another son, Murray Lowery, would also serve his country in the Navy in WWII.

A grandson, Keith Lowery, would also serve in the U.S. Marines in the 1960s and 70s. [3]

Seems there is a *strong family tradition of military service* in the Lowery family from Georgiana, Alabama.

That's not that unusual, as Alabama traditionally has been a very patriotic state, even having sent about 25 percent of the men to Texas who defended the Alamo in 1836, including its commander, William Travis of Monroe County.

LEMON O. LEE

Lemon O. Lee was a half-brother to John Chatham, husband of my mother's oldest sister Ida. Lee was born in Coffee County, Alabama, on August 15, 1895. At the time of his induction into the Army on May 24, 1918, he was a resident of Samson in Geneva County, and 22 years of age.

Lemon Lee with unknown buddy. WWI era
(date of photo ca. 1918, Source of photo: Ida Chatham family)

Archival records show he was in Company A, 322d Infantry until July 11, 1918, then part of 55 Dep. Brigade until his honorable discharge on February 1, 1919; rank was Private.[1] I did not recall meeting Lemon Lee, although I heard my father and others in the family mention him in passing. As is self-evident by the included photo, Lemon was a *Doughboy* soldier from the World War I era. Records show Lee never served overseas, as he got into service on the tail-end of World War I.

Private Lee and his unknown buddy with a Babe Ruth-like physique are in field conditions where they posed near their Army tent. They both appear to be enjoying their cigarettes, which were hot commodities among soldiers.

Cigarettes had a special place of honor among WWI soldiers, and they were highly coveted, especially among those in battle where supply lines were stretched and battle conditions often made it difficult to get commodities to those actually doing the fighting.

The use of tobacco in 1918 did not carry the stigma associated with smoking today. There were no exorbitant taxes on tobacco, nor frequent public service announcements from the government encouraging one not to smoke. Nor were there any smoke-free college campuses, restaurants, and hospital parking lots. The mantra in the World War I era was, "If you have 'em, light 'em up boys!"

When a soldier's supply ran low, "Hey buddy, can I have a drag off your smoke?" was a commonly heard request in training and in combat. Leaders of the military recognized that cigarettes for soldiers were integral to the well-being of a fighting force. Even writers and poets of that era recognized that cigarettes had a special place on the battlefield. The following poem titled "No Man Knows War" by Edwin Rolfe included a special reference:

Battle knows only three realities...enemy...rifle...life
No man knows war—or its meaning—who has not
stumbled from tree to tree, desperate for cover, or dug
his face deep in the earth, felt the ground pulse with the
ear-breaking fall of death
War...is your comrade struck dead beside you, his
shared cigarette still alive in your lips

After his military service, Lee married his first wife, Bessie. They had a son, Pate. and a daughter, Ernestine. Pate was in the U.S. Navy in WWII, and Ernestine served her country as a member of the WAAC—Women's Army Auxiliary Corps. Over 150,000 American women served in the WAAC during World War II. Members of the WAAC were the first women other than nurses to serve within the ranks of the United States Army.

ROBERT FRANK LAWRENCE

Robert Frank Lawrence was born in 1892, in Hacoda, Alabama (near Florala), to Robert "Bobby" and Sarah (Adams) Lawrence. Frank married Blanche Foxworth who died December 30, 1916, during childbirth with their daughter, Blanche.

Within nine months of Blanche's birth, Frank on September 21, 1917, at age 25, joined up with the U.S. Army for World War I (enlisting at Andalusia, Alabama, in Covington County). While he was away at war, Frank's mother, Sarah, raised the infant Blanche until Frank returned from the war.

During the war, Frank was assigned to Co. A, 106th Engineers, 31st Dixie Division. Engineers were responsible for laying pontoon bridges, using explosives to demolish objects that got in the way of the Army moving forward, clearing mines, building roads and bridges, and a host of other combat operations. His unit was at Brest, France at the time of Armistice. He was with Allied Expeditionary Forces from Sep 17, 1918 until July 5th, 1919. With the rank of Private First Class, Robert Frank Lawrence was discharged at Fort Oglethorpe, Georgia, on July 12, 1919.[1]

PFC Robert Frank Lawrence (Don Newbauer photo)

After discharge, he returned to Florala and farmed the rest of his life. He passed away in 1947, and lies in Greenwood Cemetery, in Florala, Alabama. [2]

WORLD WAR II

"Yesterday, December 7, 1941—a date which will live in infamy—the United States of America was suddenly and deliberately attacked by naval and air forces of the Empire of Japan... The attack yesterday on the Hawaiian Islands has caused severe damage to American naval and military forces. I regret to tell you that very many American lives have been lost. In addition, American ships have been reported torpedoed on the high seas between San Francisco and Honolulu. Yesterday the Japanese government also launched an attack against Malaya. Last night Japanese forces attacked Hong Kong. Last night Japanese forces attacked Guam. Last night Japanese forces attacked the Philippine Islands. Last night Japanese forces attacked Wake Island. And this morning the Japanese attacked Midway Island.

"Japan has, therefore, undertaken a surprise offensive extending throughout the Pacific area. The facts of yesterday and today speak for themselves. The people of the United States have already formed their opinions and well understand the implications to the very life and safety of our nation." [1]

So rang the words of President Franklin Roosevelt, addressing Congress the day after the attack on Pearl Harbor, when he asked the U.S. Congress to declare war.

The blow from the Pearl Harbor attack might have been somewhat lessened had authorities taken seriously an intercepted Japanese message that had been de-coded by Mrs. Dorothy Edgers of the Office of Naval Intelligence, on Saturday, December 6, 1941. She showed it to her superior officer, who said it "needed more work and could wait until Monday." [2] Other signs of the coming attack were not noticed. A Japanese spy that day walked into a Pearl Harbor telegraph office and dispatched back to Japan the locations of our Navy's warships sitting at anchor in the harbor; no mention of the telegraph was made to anyone in authority by those in the telegraph office; kind of reminds one of so many missed signs prior to the terrorist attack on 9/11/2001.

CHARLES SAUNDERS

"Handbags, driving gloves, and lamp shades"

Saunders and another MP accepting German prisoners for processing and to be put aboard a flight to Paris for imprisonment (Army photo)

Charles Saunders was born to Lawrence L. and Lena Mae (Thompkins) Saunders. He graduated from Opp High School, 1943.

In the WWII years, he went through Army Infantry training at Ft. Jackson, S.C., became a Military Policeman, then volunteered for an elite guard unit. The unit required one to have a high school diploma, a level of education that a good many soldiers of that time did not have.

Later, while assigned to Fort Custer, Michigan, he had a date with a Mary Ann Hoag in Grand Rapids. When he met her dad, he was asked by Mr. Hoag about his hometown.

He replied "Opp, Alabama, but you probably never heard of it." The father said he knew right where Opp was because he "listened to the Grand Ole Opry on Saturday nights, and (Opry star) Lew Childre had put Opp, Alabama on the map."

Saunders' outfit went ashore at Utah Beach in Normandy, France on June 12, 1944, just six days after the initial waves of the D-Day invasion.

Being in the elite guard, Saunders saw many things other soldiers did not. He often guarded the 9th Army Commander, LTG William H. Simpson. On one special occasion his unit escorted French General

Charles DeGaulle back into Paris after the Allies had declared it to be an "open city" and one not subject to bombing raids.

Other not so pleasant duties included the task of burying the dead, both German and Allied.

On another occasion, he was with a unit put on a train that bypassed Bastogne (near the center of the Battle of the Bulge) and swept around the fighting to the north and east to set up checkpoints. For heat on the trains, they used a box on the floor with sand soaked in diesel fuel, then ignited, a crude but effective way of staying warm.

When they reached a point northeast of Bastogne, they set up checkpoints—one MP in the middle of the road, one in each ditch, and a fourth posting behind as a safety watch. When German soldiers tried to sneak past the American checkpoint wearing American uniforms, they were captured one day, declared to be spies, and then put in front of a firing squad the following day. Saunders remembered vividly using leather straps to bind the condemned German soldiers' hands and legs just before they were shot.

He said it was rare that the soldiers took a bath. Most times the soldiers "stunk to high heaven," and they used their helmets to pour in water and dab wash.

During one roadblock where an MP was guiding American tanks toward Bastogne, a German sniper shot from a building and killed the American MP. A German woman ran out and told Saunders and his buddy, Ralph Ertle, she knew who did it. With her help, they ran into the building and took care of the sniper. She then inquired about her husband that she'd been told was a POW held in Mississippi. She asked if he was being treated well. Ertle replied, "He's a lot better off than we are. At least he has a warm bed at night."

Occasionally, despite the gravity of fighting a war, the soldiers had something humorous happen. Saunders' unit once was met by a German lady who yelled out, "There's a *spy* in my house." The Americans inquired as to his nationality, and she replied, "He doesn't speak French, and he doesn't speak German, and he doesn't speak English."

As they investigated, they demanded the "spy" come out of the woman's bedroom. He refused. When they barged in, he was laying on the woman's bed and had been there for some weeks. As it turned out, he was an AWOL American soldier from the hills and hollows of West Virginia who had left his unit and had been with the woman for some time.

He and his other buddies were highly amused that the lady couldn't figure out what language he spoke, but it didn't seem to have discouraged a tryst with the soldier.

On another occasion, during a burial detail, Saunders examined closely a *Luger* 9-mm pistol of a German naval officer. He remembered how impressed he was with the quality of Germany's firearms, both rifles and handguns. The Luger pistol was standard issue and a really fine handgun.

Other duties included accepting and processing surrendering German soldiers. He said German officers insisted on surrendering to an American officer of *equal or superior rank* to themselves.

The processing was time consuming, and some of those to be incarcerated seemed listless and very downtrodden. During the 1945-46 tribunals that followed the end of the war in Europe, Saunders was a guard for Germany's Hermann Goering, who founded the *Gestapo* and was head of Hitler's *Luftwaffe.* He also guarded Albert Speer, who was head of Nazi Armaments and War Production.

Saunders was in the courtroom during the Nuremberg Trials (photo below). Speer was sentenced to 20 years in prison. Goering was sentenced to hang, but ate a hidden Cyanide pill on October 15, 1946 and died before he could be hung.

Tribunal testimony was graphic. Anton Pacholegg, a prisoner at Dachau, was a clerk at Germany's "experimental station." When subjects died from torture, he said human skin was removed from their buttocks and thighs, tanned, and turned into handbags or driving gloves for German SS troops. The Commandant of Auschwitz told how he used *Cyclon B*, a crystallized prussic acid dropped into gas chambers, to kill Jews within 3-15 minutes. He said they tricked the Jews into peacefully entering the gas chambers by assuring them they were only going to be de-loused. [1] Saunders remembered testimony to the effect that the tanned human skin from torsos was also used to make lamp shades.

Another of Saunders' acquaintances, T.J. Boothe of his hometown, showed up in court at Nuremberg on one occasion. Boothe had been a draftee in early 1945, completed basic training at Fort McClellan, and was a private in the 397th MP Battallion. Boothe had only a short while earlier married Pauline Cowen, niece of Noah Cowen, and they had a young baby boy (Steve).

Hermann Goering (standing) and Rudolph Hess. Charles Saunders MP facing camera (upper left), MP Ralph Ertle right/aft shoulder of Goering

Boothe arrived Le Havre, France, then was put aboard a "40 plus 8" train (40 soldiers and 8 *cows*), and sent to Furth, Germany, adjacent Nuremberg. There, he was assigned as a cook for the American Army officers' mess.

When offered the opportunity to attend the trials, Boothe jumped at the chance in June, 1946, and told this author he witnessed the trial of a Nazi who had converted a church in the Netherlands into a training facility for German soldiers after the Germans occupied the area years earlier. [2]

After the war in Europe ended, Saunders' MP company was sent to the Pacific Theater. They were put on a ship as part of the Army's 81st Infantry Division, but never made it all the way. After they sailed, Japan surrendered, and the ship turned around. They were on the water 38 days. Saunders remained in Europe for the Nuremberg Trials.

Charles Saunders' brother, Robert Lawrence Saunders, joined the Marine Corps. Charles recalled both he and Robert made ample use of the many benefits of the G.I. Bill so they could go to college. Without it, they likely would never have been able to pay tuition. That was also true of thousands of other young Americans in that era.

Unlike Charles, Robert missed the action in WWII, as he was in Officer Candidate School, but was then sent to Okinawa to prepare for invading Japan, which never happened because the Japanese surrendered.

Robert went on into Japan as part of the Occupation Forces, and likely was exposed to radiation by being there so soon after the A-bombs were dropped.

When the Korean War broke out, Robert was called up and did participate. Robert at one time was principal at Tallassee High School. He later was a Dean at Memphis State University and an Assistant Dean at Auburn University.

Robert also had another unique experience while in the Marine Corps. He was one of 50 volunteers who observed the explosion of an Atomic Bomb during testing at *Frenchman's Flats*, Nevada, using eye goggles with a special covered lens, from a distance of more than 10 miles away. Even that far away, they still felt the immense heat of the detonation, and he no doubt was exposed to high radiation.

According to his wife, Dot, Robert was always classified as an "Atomic Veteran" whose health was tracked by the Defense Department. Atomic Veterans received special priority enrollment for health care services from VA for radiation-related conditions. [3] Robert died of cancer in 2002.

After discharge from the Army at Camp Shelby, Mississippi, Charles Saunders married Marion Livingston in 1948, and they raised three daughters—Beth, Nancy and Susan. Susan passed away at age 28 from ALS (Lou Gehrig's disease) which was very heartbreaking to him and his family and friends.

After taking off his uniform, he earned an undergraduate degree from Troy State, then worked successfully to get a Master's Degree from Auburn University (one of his fellow graduates was Wayne Teague who later was prominent in Alabama education circles). Saunders at one time was Principal of A.G. Parrish High School in Selma, then later was Principal at Central High School in Phenix City, Alabama.

In 1970, Charles got a Specialist in Education degree from Auburn University. That same year, Saunders was brought to Troy specifically to integrate Charles Henderson High School and Troy's city schools. Meeting resistance from the local school board, he was personally brought to Federal Judge Frank M. Johnson's chambers in Montgomery in November, 1971, and told to get with the program.

Troy's Board of Education had a plan to send Troy male students to the all-black high school in Troy, and send all Troy female students to Charles Henderson High School. Judge Johnson refused to go along, and

Troy's schools ultimately had to comply with Johnson's ruling and fully integrate all the schools.

When Saunders asked Judge Johnson what would happen if he couldn't get the board's agreement on integrating, Johnson replied, "I'll hold you in contempt."

Knowing the answer before he posed a question, Saunders then asked, "What does that mean, judge?" Johnson replied, "It means I'll throw you in jail for a very long time."

Johnson wanted it done in one day's time, which was totally impossible. However, Johnson did give Saunders until January 4, 1972, to get it done. Saunders returned to Troy and passed along the word. Troy's city schools were subsequently integrated in accordance with Judge Johnson's guidance.

Saunders later worked for the Alabama State Department of Education where he rated different schools and their accreditation. He worked under Dr. Wayne Teague who was state director at that time. [4]

Having served his country so well in the military and in public education, he now enjoys the retired life and lives in Troy near Charles Henderson High School on Crowe Hill.

Charles Saunders' wife, Marion, passed away in 2009, but he keeps very busy answering requests to appear at public events and talk with others about historical events.

Saunders is in demand as a public speaker and travels extensively to share with others his life experiences, which are very unique.

RICHARD E. COLE

Author with Richard Cole, age 97, Air Commando Association banquet, Ft. Walton Beach, Florida, Oct. 12, 2012 (Personal photo)

On Sunday, April 19, 1942, when the earliest rays of dawn sunlight first peeked through their windows, residents of Columbus, Georgia, stirred and went outside to pick up their morning papers.

The headlines of the Columbus *Sunday Ledger-Inquirer* made sure they got their 10 cents worth of news that morning. In bold letters, the headlines read: GIGANTIC OFFENSIVE AGAINST JAPS IS BELIEVED NEAR. Under the headline, "U.S. Airmen Thought To Have Inflicted Great Loss In Raid" was the caption of an article by AP War Editor Richard McMurray.

McMurray wrote "Tokyo and three other great Japanese cities shuddered yesterday from the destructive crash of United States bombs in what may have been the start of an offensive..."

Local residents at the time had no idea who those gallant airmen were who flew their B-25 bombers off an aircraft carrier in the Pacific and attacked Japan. But, news has a way of getting out, and they would soon learn of the exploits of those airmen, one of whom was Richard Cole.

Lieutenant Colonel Richard E. Cole, USAF (retired), was born in 1915, in Dayton, Ohio. Cole entered the Army Air Corps in 1940 and received his aviation wings in 1941. He was immortalized in American aviation history as a young lieutenant while flying as co-pilot with then-Lieutenant Colonel Jimmy Doolittle on their daring mission flying B-25 bombers on April 18, 1942, to strike mainland Japan.

The group came to be known throughout history as Doolittle's *Tokyo Raiders.* Eventually, all the men on those raids would be enshrined in the American psyche through movies such as *Thirty Seconds Over Tokyo.*

As of this writing, only four survive, and one is Cole that I met in Florida in October, 2012, and again in April 2013. The other three are retired Lt Col Robert L. Hite of crew #16 (who spent 40 months in a Japanese prison), S/Sgt David J. Thatcher who was an engineer-gunner on crew #7, and Lt Col Edward J. Saylor who was the engineer on crew #15.

President Franklin Roosevelt wanted to deliver a retaliatory strike from America that would show the world that, despite the crippling blow Japan dealt the U.S. Naval Fleet at Pearl Harbor some four and one-half months earlier, America was a force to be reckoned with.

The strike had other purposes, one of which was to cause the Japanese to doubt their leaders, and the other to boost the spirits of the American people and the morale of America's fighting men and women.

The raid had immediate repercussions among the Japanese Empire. It made them realize that they were not invincible, and it caused their military leaders to keep many of their forces on and around the home island in order to protect the homeland. They also re-thought their strategy in the Pacific, with many more Japanese then agreeing with Admiral Yamamoto that Midway Island had to be captured and the American Naval Fleet destroyed. That decision and change in strategy would ultimately lead to their defeat in the Battle of Midway.

A call for volunteers went out, with a very positive response. Crews were picked, and rehearsal flight training for the raid was conducted for three weeks beginning March 1, 1942, at Eglin AFB, Florida, on Auxiliary Field 1. A Navy flight officer, Lt Henry L. Miller, from Pensacola, taught the aircrews techniques for launching off an aircraft carrier.

I later learned that Lt Miller married the former Lucille Dean of my hometown of Opp, Alabama.[1] The Deans had a pharmacy and soda shop, and Miller's nephew, Rego Rue, remembered that Lt Miller occasionally loaded up Lucille in a small aircraft and buzzed his Granny Dean's house on East Ida Avenue. He also recalled that their son, young Henry "Panky" Miller, Jr., came to town during summer months to spend time with Rego and his friend Butch Bracke. They spent a lot of time in the city pool in the summers.

Lt Miller training the pilots was just one part of the raid planning. Many modifications had to be made to the B-25s. "The sixteen planes the Doolittle Raiders would use had to be modified so they could carry a ton of bombs and fly as far as twenty-four-hundred miles. Their B-25s got extra gas tanks inside the fuselage, new bomb shackles, specially-adjusted engines, and broomsticks painted black to simulate fifty-caliber tail guns. Their Norden bombsights were removed—(they were of) no use at low altitude. The Raiders' armament officer, Captain Ross Greening, devised a twenty-cent gizmo out of two pieces of aluminum, and it worked just fine at fifteen hundred feet." [2]

Doolittle's crew made their departure run off the Navy carrier USS *Hornet* on aircraft number 40-2344. A Japanese boat had spotted the carrier with the aircraft aboard as it steamed toward Japan, so Doolittle had to make the decision to launch the crews far earlier than originally planned. The early launch meant they had less fuel left over to make their intended landings at airfields around Chuchow (China), about 100 miles inland from the Japanese-controlled coastline, but that didn't deter Doolittle.

Sailors aboard the *Hornet* took bets on whether the B-25s could get airborne off the ship's runway. There were skeptics. "A Navy pilot shouted to anyone in earshot 'He won't make it! He can't make it!' But, as Jimmy's navigator, Hank Potter, later recalled, 'We were particularly confident since we had the best pilot in the Air Force flying with us.'" [3]

The bombardier on the aircraft with Doolittle, Cole, and Potter was Staff Sergeant Fred A. Braemer, and the gunner was Staff Sergeant Paul J. Leonard. Targets included industrial areas of Tokyo, Yokohama, Nagoya, Osaka, and Kobe, as well as the Yokosuka Naval Base.

The crews completed their bombing runs, and Doolittle's crew continued on toward China for several hours flying time in rain and poor weather. Cole stated that, "When the low-fuel lights came on in the cockpit, the crew got up from their crew positions and parachuted out" of their B-25.

He said he landed in a tree in darkness, with his parachute canopy and risers entangled in the tree limbs. Not knowing how far off the ground he was dangling, Cole opted to wait for daylight to cut himself loose and climb down (from about 10 feet off the ground).

The Chinese people and military forces operating in opposition to the occupying Japanese helped him get back to friendly American forces. He stayed in the China-Burma-India theater until 1943, continuing to fly and support the war effort. Cole eventually retired as an Air Force command pilot and achieved the rank of lieutenant colonel. [4]

Doolittle's aircrews were lauded for their daring airmanship in carrying out a mission so critical to our nation. Lieutenant Colonel Doolittle was quickly promoted to brigadier general, skipping the rank of colonel, and was later awarded the Medal of Honor by President Roosevelt for planning such a long-shot mission and leading his men through such hazardous bomb runs over hostile territory.

Ironically, Cole said Doolittle himself had thought the mission would probably be dubbed a failure because all the aircraft were lost and a good number of crewmen were injured, killed, imprisoned, or executed. A grateful president and nation thought otherwise.

Cole said three of the captured American airmen were executed by the Japanese who claimed they were war criminals, accusing them of strafing civilians; another died in captivity. Others were held captive in Russia after landing not far from Vladivostok.

With the exception of the Wright Brothers, Charles Lindbergh, and Eddie Rickenbacker, Jimmy Doolittle may be the most recognizable aviator name in American history. In the WWII era, his name was on the tip of every adult's tongue after he planned and led the daring raid on mainland Japan in April, 1942.

As a military leader, he was aggressive in taking the fight to the enemy. After the Tokyo Raid and his assuming rank of brigadier general, he filled many command positions, including Twelfth Air Force in North Africa and later Fifteenth Air Force in the Mediterranean Theater of Operations.

He wanted to continue flying bombing missions himself, but his higher-ups ruled that too dangerous because of his knowledge of *Ultra,* which could be compromised, should he be shot down, captured and interrogated *(Ultra* was the allied program that had various successes in breaking the codes of German communications).

From January 1944 to September 1945, Doolittle held his largest command, the Eighth Air Force (8 AF) in England as a lieutenant general, his promotion date being March 13, 1944, and the highest rank ever held by a Reserve officer in modern times.

He was placed in this command after General Henry "Hap" Arnold became dissatisfied with the way General Ira Eaker was prosecuting the air war in Europe (Eaker was too conservative on utilizing the capabilities of his fighter escorts, in Arnold's mind).

Doolittle's major influence on the European air war occurred early in the year when he changed the policy requiring escorting fighters to remain with the bombers at all times.

With his permission, P-38s, P-47s, and P-51s on escort missions strafed German airfields and transport while returning to base, contributing significantly to the achievement of air supremacy by Allied Air Forces over Europe. [5]

Retired Lt Col Richard Cole has five children, daughters Cindy (Cole) Chal and Chris, as well as sons Richard, Sam, and Andrew. His son, Richard, became an accomplished Air Force pilot, flying the F-15 *Eagle*. Cindy has one son, Nathan, who flies Air Force KC-135 tanker aircraft. Another of Cindy's sons, Elliot, is a cadet in his second year at West Point. [6]

After his Air Force career was complete, Cole and wife Martha retired to the area near San Antonio, Texas, where he was a custom home builder for many years. He is still active, cutting his own fire wood and attending events honoring the heroics of the Doolittle Raiders.

When Richard Cole flew that Tokyo Raid on April 18, 1942, with Jimmy Doolittle, he was in the cockpit with a man that would ultimately change the course of the war in the Pacific and in Europe, helping to bring WWII to a quicker, more successful conclusion by 1945. Richard Cole was part of a history-changing military event critical to our nation.

He has nothing but praise for Jimmy Doolittle, recalling that during preparations for the mission, Doolittle was usually the last man off the flight line. He always was around to make sure the crew chiefs serviced and refueled the aircraft, and made sure they had a ride back to their quarters. A true leader does things such as that.

I was honored to meet Colonel Cole in October, 2012 and April, 2013. I'm honored to salute him, as well as all the members of his family who have and are currently contributing so much to the defense of our country.

JAMES RALPH "SHUG" JORDAN

"He asked to be stitched up so he could stay with his troops."

Coach Ralph "Shug" Jordan

"You're so right, Carl, you are so right." On Sunday afternoon televised playbacks of Auburn University football games, with sportscaster/commentator Carl Stephens hosting Auburn football coach Ralph "Shug" Jordan, this became one of the most-often-quoted comments in all of SEC football history.

It was common knowledge that Shug Jordan was a great football coach, as evidenced by his 176 wins against 83 losses and 6 ties at Auburn. What was not common knowledge to viewers was the outstanding war record of Coach Jordan during his time in the Army in World War II.

Born in Selma, Alabama, September 25, 1910, he was nicknamed Shug because of his love for sugar cane. He was married to the former Evelyn Walker of Columbia, South Carolina, whom he met in 1934, and they have two children, Susan Dalby and Ralph, Jr. [1]

During World War II Shug Jordan was in the first wave of troops in the invasions of North Africa, Sicily, and Normandy (on D-Day, June 6, 1944), where he was wounded and awarded the Purple Heart and the Bronze Star. Recovering from his wounds, he participated as a front line officer in the invasion of Okinawa and in the Philippines campaign.

Guest columnist Alvin Benn, writing for the *Montgomery Advertiser* on June 6, 2010, the 66th anniversary of D-Day, had the following to say about Shug Jordan's war service:

"This is the day 'Shug' almost died. It was June 6, 1944 and Army Captain Ralph Jordan had just splashed ashore at Normandy to lead his engineers in setting explosives to penetrate a German seawall aimed at slowing the Allied advance.

It was imperative for the (combat) engineers to do their job quickly because tanks were waiting on the beach for the seawall to be breached in order for them to push through toward the French interior.

Moments later, a German 88-mm artillery shell exploded nearby on Utah Beach, sending shrapnel into his left shoulder and showering others in his brigade. He might have died that morning if an artery had been severed by the shrapnel, but luck was with him.

Other troops with him were killed or seriously wounded on the first day of the D-Day invasion.

Most college football fans know 'Shug' Jordan as a coaching legend who guided the Auburn Tigers to a national football championship in 1957 and had more victories than any gridiron coach in the school's history.

What they most likely don't know is the fact that he was also quite a soldier, a man who took part in much more than the Normandy invasion 66 years ago today.

Jordan's military record is well documented, but he tended to downplay his role, and when he came home he never made a big thing out of his Purple Heart for that wound at Normandy. The same applied for other combat medals and ribbons that filled several rows of his uniform for action during invasions of North Africa, Sicily, Italy, Normandy, and Okinawa, which was halfway around the world. It could be easily assumed that few U.S. soldiers took part in five amphibious invasions from Europe to the South Pacific during World War II. Shug' was one who did."[2]

James Ralph Jordan returned to his alma mater as head coach of the Auburn football team in 1951. He had graduated from Alabama Polytechnic Institute in 1932, lettering in football, baseball, and basketball. By a vote of his teammates he was designated Auburn's Most Outstanding Athlete in 1931-32.[3]

Alvin Benn wrote further: *"Author Rich Donnell was one of the last journalists to interview Jordan, who died 30 years ago next month. In* SHUG: The Life and Times of Auburn's Ralph 'Shug' Jordan, *the coach talked about the war and what he saw around him at Utah Beach. 'I saw 'em stacked up like cordwood,' he told Donnell. 'I could easily have been one of them. I had been scared before I went over there, and I was damn sure scared over there.'*

Captain Henry McHarg, another officer in Jordan's engineer brigade, wrote Auburn University President Hanly Funderburk about his friend's courage under fire and after being wounded.

Jordan refused to be evacuated for treatment, McHarg said. He basically asked to be stitched up so he could stay with his troops during the initial phase of the invasion. Ralph Jordan Jr. said he learned that after his father was wounded, he and a second soldier could not be found. That's when a search party was formed to look for them. 'They found my dad propped up against a seawall smoking a cigarette,' Jordan Jr. said during a telephone interview from his home near Knoxville, Tenn. 'He was waiting for help to arrive.'" [4]

Returning to Auburn in 1945, he coached for one year before joining the coaching staff of the professional Miami Seahawks. Mid-season, 1946, he became an assistant coach at the University of Georgia.

After a series of losing seasons for Auburn, Coach Jordan was made head coach and produced a 5-5 record for his first year. At the end of six years he had built an outstanding team which won both Southeastern Conference and National Championships.

Consistently producing good football teams, Coach Jordan was named SEC "Coach of the Year" in 1953, 1957, and 1963. During eight different seasons he was selected national "Coach of the Week," and his Auburn Tigers were invited to play in eight post-seasonal bowl games.[5]

TUPPIE LEROY RUSHING

"Demolitions Rock Manila. Japs Cornered. Encircled Nipponese Commit Atrocities Against Helpless."

Corporal Tuppie Leroy Rushing in uniform 3

Corporal Tuppie Leroy Rushing was born February 23, 1919, to Tup and Maryann "Minnie" (Floyd) Rushing in Walton County, Florida (Panhandle area). Minnie was my maternal grandmother, Hettie's sister.

As a young man, Tuppie worked as an electrician for the Rural Electrification Administration (REA—one of Franklin Roosevelt's *New Deal* programs), as well as the Civilian Conservation Corps—or CCC (another New Deal program).

In his leisure time, he hunted raccoon and fished for catfish and bream in the Choctawhatchee River near his home.

He married his sweetheart, Katie Mae Wilson, in April, 1940. Pearl Harbor was attacked on December 7, 1941, and their first child was on the way when he was drafted into the U.S. Army on 24 February 1943, just one day after his 24th birthday. That child, Anita Mae, was born on October 15th that year as Tuppie was training and gearing up for a war that would be fought by *America's Greatest Generation.* [1]

Following in the tradition of his cousin, Will Rushing, who had served so well in WWI, Tuppie was in the Infantry in the WWII era. He was inducted 17 Feb 43 at Camp Blanding, Florida, near Jacksonville. Rushing had been an arc welder as a civilian, so he had some skill beforehand, having worked a year at the Wainwright Shipyard in Panama City, Florida. His Company D, 14th Battalion CO at Ft. McClellan, AL was 1st Lt P.M. Henry, Jr. After basic training, Rushing was made a cook, and later a barber to cut hair of other troops. On a ship en-route to Guam, he stayed seasick a lot, a common occurrence for troops.

In the Army, he was a Field Lineman, specialty 641, the job description of which read, "Laid, maintained, and salvaged telephone or telegraph wire. Ran wire along the ground by guiding it from a wire reel of a wire truck. Fastened wire to poles, stakes or trees. Located and corrected shorts and opens by splicing breaks and cutting out defective wire." His separation papers showed he served in the South Pacific, Philippines, and Japan.

His unit, Co. I, 4th Infantry Regiment, landed in Manila, Philippine Islands in 1944, the morning after the first wave of the invasion by American forces had begun.

The headlines of the *Florence* (Alabama) *Times* rang out with descriptions of the heroic action of the Allies in Manila that day: "Manila (AP). Demolitions Rock Manila. Japs Cornered. Encircled Nipponese Commit Atrocities Against Helpless." The AP article continued: "Under skies black from new enemy demolition destruction, American troops cautiously tightened an inevitable death noose today on Japanese so desperately that they shot and bayoneted civilians in south Manila." [2]

A crack shot and a "tough cookie," Tuppie Rushing fought admirably with the American Army against a seasoned Japanese occupation force.

He told once of having Japanese soldiers sneak up in the dark on a two-man foxhole of his and a buddy's and taking a shot that narrowly missed. He didn't sleep too well in foxholes the rest of the time in the war.

The Manila victory was sweet revenge for American General Douglas MacArthur who had to evacuate from Corregidor to Australia when the Japanese first invaded the Philippines in 1941, occupied the Islands, and forced Lieutenant General Jonathan Wainwright's captured American soldiers on what came to be known as "The Bataan Death March," with the resultant death of several thousand American military men. General Wainwright was among those captured and later held as POWs in Manchuria.

MacArthur had issued a very public vow of "I Shall Return," and he did, *with a vengeance*, aided by the bravery of those such as Tuppie Leroy Rushing from the Florida Panhandle area, a second cousin of this author, and one whose service to the nation I am extremely proud to salute.

Along the way, he earned numerous awards and decorations, to include the Combat Infantry Badge, Good Conduct Medal, MKM Rifle, Philippine Liberation, CIR 136 WD 45 w/Star, APTO Medal, Good Conduct Medal, and World War II Victory Medal.

Following the surrender of Japan after the dropping of Atomic bombs on Hiroshima and Nagasaki, Tuppie Rushing was discharged from the Army 10 January 1946, at Camp Blanding, after attaining the rank of corporal. He had 10 months and 24 days of continental service, as well as two years of overseas service.

After returning home, he and Katie reunited and raised Anita and Margie, and three sons in their Walton County home. Tuppie continued to love hunting and fishing, but he never would go offshore deep-sea fishing. All that sea sickness on the way to Guam could never be forgotten. As long as he was in sight of land, he liked to fish, but would never go out of sight of land. [4]

Three of the five children were boys—Eddie Lee, Chester Leon, and Kenny Earl Rushing. Tuppie passed away in 2007 at age 88.

All three of the boys later served in the U.S. Army to continue on with the time-honored tradition of the Rushing family in supporting their country in its time of need.

Two of the three served in the Vietnam War, which is another story all in its own.

ARTHUR ROBERT SCROGGINS

Scroggins briefing his aircrew (his personal photo)

Arthur Robert Scroggins was born February 6, 1922, to parents Arthur and Ruby (Durden) Scroggins in Tallassee, Alabama. Robert's father for many years ran a furniture store in his hometown.

He and future wife, Irene Houston, were watching a movie in the Paramount Theater on December 7, 1941, when an announcement was made that told of the Japanese bombing of America's Navy at Pearl Harbor. The next day he was at the recruiter volunteering for duty. He also remembered vividly the sight of anti-aircraft machine guns that were quickly placed at the new power generating dam crossing the Tallapoosa River in Tallassee.

In early 1942, at age 19, he and two of his high school classmates tested for the Army Air Corps. He was the only one of the three who passed. His buddies wanted him to go with them and enlist in the Marine Corps, but he decided he'd stick with his original plan of becoming an aviator.

Scroggins qualified as a Bombardier aboard B-17s in World War II and had numerous close calls and narrow escapes while flying 34 combat missions over Germany. He was in the 335th Bomb Squadron, 95th Bomb Group, which was part of 8th Air Force. He spent time in the 336th HQ Sq. as well.

In June, 1943, he married Irene, whose family was originally from Garland, Alabama (near Georgiana), but had moved to Coosada when she was age nine (they celebrate their 71st anniversary in 2014).

One crash Robert survived was in the English Channel on January 4, 1944, after his shot up and crippled B-17 *Superstitious Aloysius*, piloted by Peter V. Milward and Arthur Farris, Jr., went down in extremely cold waters. The British described conditions that day as, "Bitterly cold with a freshening wind."

He and crew were lucky as there was a Royal Navy destroyer in the vicinity that plucked the distressed Americans from the frigid waters.

The HMS *Verdun* was a coal-fired, steam-driven, destroyer leftover from WWI, that was on East Coast convoy escort duty accompanying merchant ships that saved them that day. The *Verdun* fired a line that the ditched crew couldn't reach. The ship turned and fired another line which the crew was then able to get to, so they were rescued.

It was so cold that their uniforms were iced, so their rescuers took knives and cut them out of the frozen outfits, then dressed them in British uniforms in which to warm. To thaw out the survivors, they poured room-temperature water on them and tried to force some good rum down their throats.

At a mess hall, once they reached dry land, a fellow asked Robert if he had a brother in the American Air Force, because he looked just like a man he had been seeing in an American uniform!

Scroggins related the story of another close call that occurred in July, 1944, when he was placed in a P-38 piloted by Colonel Hubert A. "Hub" Zemke that crash landed back in England after being hit by enemy anti-aircraft fire. The P-38 had been specially modified, with guns removed, a Norden bombsight installed, a panel removed, and Scroggins crammed down in the space with instructions to use the bombsight and give a "10-9-8-7-6-5-4-3-2-1" countdown to drop bombs. With the panel re-secured and no way to bailout, Scroggins wore no parachute. Zemke, an experienced pilot from fighting in

China, wanted to experiment to see if he could lead a flight of fighter bombers to make drops on this specially-configured P-38's command.

When they departed England on this flight, the idea was that Scroggins would be aboard this P-38 piloted by Zemke and command the drop of two 500-pound bombs on a target at Montdidier, France—north of Paris—that Zimke had spotted two days before. The target consisted of four JU-88 German aircraft that had been pushed up under trees for concealment.

Scroggins said General Jimmy Doolittle praised Zemke as his "greatest fighter group commander." Zemke had shot down a German ME-262 jet and led the first bomber escort mission over Berlin. He eventually was shot down and became a POW responsible for 7,000 other prisoners at *Stalag Luft I* in Barth, Germany, after having a wing come off his P-51 and bailing out. He commanded the 56th Fighter Group in England, which came to be known as *Zemke's Wolfpack.* [1]

Enroute to the target, their P-38 was hit by German anti-aircraft fire which knocked off the right side propeller and its spinner, and components of the right engine. The other engine was seriously leaking oil from shrapnel, causing them to crash land back in England because they had no brakes and ran off the end of the runway. With lady luck on his side, Scroggins miraculously survived the one-time bombing mission with Zemke.

Scroggins' next close call was in August, 1944, aboard a B-17 flying a mission on which he suffered his most serious wounds. They were bombing the bridges of Cologne, Germany, when an enemy anti-aircraft round struck the nose of his aircraft, killing the navigator and knocking him over, and throwing Scroggins across the dead crewmember. They did a controlled crash on an airfield in Brussels, Belgium.

During the Bombing of Cologne in World War II, Cologne endured 262 air raids by the Western Allies, which caused approximately 20,000 civilian casualties and almost completely wiped out the centre of the city. During the night of 31 May 1942, Cologne was the target of "Operation Millennium," the first 1,000-bomber raid by the Royal Air Force in World War II. 1,046 heavy bombers attacked their target with 1,455 tons of explosives, approximately two-thirds of which were incendiary. This raid lasted about 75 minutes, destroyed 600 acres (243 ha) of built-up area, killed 486 civilians and made 59,000 people homeless. By the end of the war, the population of

Cologne had been reduced by 95 percent. This loss was mainly caused by a massive evacuation of the people to more rural areas.[2]

When he was brought to a British hospital, his surgeon, Dr. Sam Upchurch, asked him where he was from as he was wheeled into the operating room. He replied, "Tallassee, a little town in Alabama about 30 miles east of Montgomery that you've probably never heard of." The doctor informed him he knew exactly where Tallassee was, as he himself was a native of Clanton. Many decades later, they had a reunion after Upchurch established a highly successful plastic surgery practice in Birmingham.

Scroggins crew also landed in Russia, on occasion. He mentioned that the Germans frequently dropped sabotage contraptions on their runways that always had a spike facing upward, in an effort to damage the American aircraft wheels and hopefully cartwheel them on their takeoffs and landings.

After his surgery, he was back home on a 30-day leave, but then volunteered to go back to his unit and back to combat. On one occasion, his aircraft was attacked by a German jet fighter, the first time any of his aircrew had ever seen a jet.

At the end of the war, his crew flew humanitarian missions, dropping bundles of food to starving Dutch citizens. By that time there existed an agreement with the Germans that they would not fire on humanitarian aircraft.

After the end of World War II, he was assigned to a base in Sioux Falls, S.D., where the commander had to dream up things for the troops to do such as hours-on-end parade ground drills. He decided if that was all there was to do in the peacetime Air Corps, it was time to hang it up, so he left active duty and returned home.

Later, he was contacted by a friend who was establishing a Rifle Company in the Alabama National Guard. The friend was to occupy the company commander slot as a captain, but he offered Scroggins the deputy position as a first lieutenant, which meant a decrease in rank.

He accepted the offer and went to Fort Benning, Georgia, for Infantry training. During the Korean War era, he spent two years training others at Camp Atterbury, Indiana, and Fort Jackson, South Carolina.

Along the way he left the National Guard, went to work for the U.S. Postal Service (on January 1, 1947) and became the Postmaster in Tallassee. He worked some 30 years with the Postal Service before retiring.

To get enough "good time" to qualify for a retirement from the military, he went back into the National Guard as an E-5 (Sergeant). He was testing for Warrant Officer the day that John F. Kennedy was assassinated, an event that he says he will never forget.

Among his awards and decorations are the Purple Heart, Victory Medal, Air Medal (5 Oak Leaf Clusters), and Distinguished Flying Cross (with 1 Oak Leaf Cluster), plus others.

Scroggins is proud of his military service, and proud of his ancestors' service as well. A great-grandfather (named Harrelson) was a captain in the Confederate States Army and walked all the way from Illinois to Alabama at the conclusion of the Civil War.

One of his grandsons is a crew chief on F-16 fighters in the Alabama Air National Guard. Robert swore in his grandson with his oath of office, and recently gave this grandson ownership of his World War II uniforms for safekeeping.

The grandson, Technical Sergeant Shane Scroggins, has named his F-16 of the Alabama National Guard the *Superstitious Aloysious II* in honor of Robert's WWII B-17. No doubt the F-16 pilots are proud to fly this fighter aircraft. And, they should be, for it has a wonderful legacy, thanks to those who crewed aboard that B-17 such as Arthur Robert Scroggins.[3]

RALPH ODELL WARD

Ralph Odell Ward with brother Julian (Ralph's 2 Apr 1952 photo)

Ralph Odell Ward was born July 21, 1924, in the little town of Hosford, about 20 miles south of Tallahassee, Florida. His parents were Rufus and Cora (Lassitter) Ward.

Ralph's great-great grandfather, Brinkley Lassitter, had migrated to the new world, arriving in Savannah as an indentured servant to work seven years to clear a debt he had incurred in his home country.

Brinkley married a Creek Indian squaw who was under threat of forcible removal on the *Trail of Tears* to the Oklahoma Territories. To avoid her deportation, he headed to Florida in 1825 where members of a Creek tribe built a house for him and his bride and helped avoid the deportation. The Lassiter family grew and prospered in Florida. My Aunt Lula Rushing married a John M. Lassitter (in Ralph's family).

In his youth, at a time when all his friends called him Odell, Ward related in my interview that he had known the family of my Uncle Will Rushing who had fought in WWI (as described earlier). Uncle Will and Aunt Addie had a son and some girls that Ward was attracted to, in particular one named Wadine (nickname "Punk").

When the Rushing family had a social in their home, Ralph and his second cousin, Paul, would navigate their little motor boat up the Choctawhatchee River to the Red Bay landing, disembark and attend the social, in the faint hope that he "could hold Punk's hand and walk with her."

He described her as "so pretty with curly brown hair and features that would attract any warm-blooded young man." Alas, he didn't have much luck currying favor with Punk because his cousin, Arad "Stanley" Ward, was Punk's favorite, and Stanley also "stuck to her like glue," all the while keeping a close eye on any attentions that Ralph and Paul might try to shower on Punk. Stanley and Punk were eventually married.

After a failed short first marriage, Ralph later after WWII married a girl named Lillian Sansbury from Daleville, Alabama. Together, they raised four children, and their marriage lasted almost 60 years until she passed away after a brief illness. One of their sons died in an auto accident.

In his lifetime of service, Ward was first in the CCC in Idaho where he taught young men to read and write. He later was in the U.S. Coast Guard, U.S. Army, U.S. Army Air Corps, and later the U.S. Air Force. He volunteered for duty in the Coast Guard in 1942 at age 17, shortly after the United States entered World War II and was trained to operate the guns on a Landing Ship—Tank (or LST) used to transport men and equipment to the beachheads during wartime.

The Coast Guard operated as part of the U.S. Navy, and Ward was aboard LST-207 during the reinforcements of Guadalcanal and Okinawa during the Pacific Campaign.

The 328- x 50-foot vessel carried troops and military equipment, displaced 2,100 tons at loaded weight, and cruised at 9 knots with a range of 24,000 miles. The vessel had lots of armament for self-protection: 7 x 40-mm guns, 6 x 20-mm guns, 2 x .50 caliber guns, and 4 x .30 caliber guns.[1]

On one trip, according to LST-207 logs, she arrived Guadalcanal (east of New Guinea) on 22 October, 1943 and offloaded Army and Navy personnel and cargo. Our First Marine Division (Reinforced) had first landed there on 7 August, 1941, established a beachhead, and held in the early stages of rolling back the Empire of Japan.

In early November, the LST crew began preparing for the operations at Bougainville, departing for Bougainville on 4 November and arriving Puruato Island, Bougainville, on 6 November with Marine supplies. The following is a typical account of some of the exciting action aboard Ward's vessel:

"She returned to Guadalcanal on the 8th and on the 15th started again for Bougainville. On the 17th a large column of black smoke was observed off her bow after two explosions were felt and an APD in the convoy was found to have been torpedoed.

"At 0337 enemy aircraft were sighted about 1,400 yards distant slightly aft the starboard beam and low. The 207 opened fire and after 300 rounds the plane, identified as a Mitsubishi G4M 'Betty' medium bomber, burst into flames, passed over the vessel, veered off to the right and crashed into the water two points off the LST's port bow at a distance of 300 yards.

"The commanding officer of another LST saw a torpedo wake cross LST-207's bow just previous to her opening fire. At 0342 a second plane crossed the stern from starboard to port about 800 yards distant and then veered to left away from, but parallel to the LSTs track. The 207 opened fire. Other ships astern also opened fire and the plane crashed into the water, burned and exploded.

"At 0345 the ship fired at another plane with unknown results. At 0356 a plane coming in low passed the length of the ship, distance 300 yards, and all guns, including those of nearby vessels fired with unknown results. Later that morning the LST beached at Puruato and began unloading U. S. Army personnel." [2]

LST-207 and crew was also involved in the operations at Kwajalein in June, 1944, the invasion at Guam, and the invasion of Okinawa from 1-11 May, 1945. The ship and its crew saw a lot of action in General MacArthur's Pacific island-hopping campaign that led to victory over Japan.

The 20-mm had high-explosive (HE) rounds in a machine gun that was prone to jam occasionally, thus creating a dangerous situation should some of the HE rounds jam and cook off and explode, throwing shrapnel around. Ralph was responsible for two of the six 20-mm, plus one 40-mm gun.

Ward described his top priority as a gunner: "If the 20-mm gun jammed, it was my job to run and remove the barrel and throw it overboard. There was not a second to waste."

Toward the end of WWII their single-barrel 40-mm was replaced by a 4-barrel "quad-40" radar controlled weapon that was "a thousand times better than the older gun." In his words, "If the crew could see an airplane, they could likely shoot it down." He later was sent to a school in California to learn to operate a new 5-inch gun which was very noisy. There was a "blabber mouth" kid who was always mouthing off about how he could handle most any gun far better than all the other trainees. When the kid was placed on the new 5-inch and it began firing away, he passed out from fright. From that moment on, not a single smart-aleck comment was heard from "blabber-mouth."

During one landing, Ward took shrapnel through the neck and ended up in a hospital in California.

After getting all the excitement he could handle aboard the LST-207, Ward later joined the Army Air Corps and qualified as a crew-member aboard B-25 Bombers. He also flew the venerable C-47 cargo plane. He was in B-25 training preparing for the invasion of the Japanese mainland, but the crews had no orders to indicate what the mission entailed.

After the 1945 atomic bombs were dropped on Hiroshima and Nagasaki (August 6 and August 9, respectively), Japan agreed to an unconditional surrender. It was only after their surrender that Ralph Ward and his fellow airmen were shown the orders that would have taken them over Japan had there been no surrender.

Estimates of the number of American casualties expected in an invasion of mainland Japan ran high—some said one million, others estimated far more, had the surrender not taken place.

Ward flew B-25s in other locations such as Saudi Arabia and Iran, but is not at liberty to discuss those operations without compromising information he believes may still be classified.

He also flew C-47 cargo aircraft. One memorable series of missions involved flights from Eglin AFB, Florida, to Wright-Patterson AFB, Ohio to backhaul a 12,000-pound load of liquid oxygen (LOX), which was an extremely explosive cargo. At the time, he was young and didn't think too much about the dangers. He "felt 10 feet tall and bullet proof" as young men often do in their early years.

He was under the impression the LOX was used in the launch of missiles from Eglin out over the Gulf of Mexico (at the time Eglin was known to launch BOMARC missiles).

He noted that when they taxied out with their dangerous load at Wright-Patterson, a fire truck and ambulance would shadow them through their ground taxi and takeoff roll until well after they were airborne.

Upon arrival Eglin AFB, the reverse was true. A fire truck and ambulance would run alongside the aircraft during its landing rollout. Rules did not allow them to fly higher than 10,000 feet because of the increased danger of explosion of their cargo.

At the time, he was flying as a co-pilot for pilot David Raiford who always handed him the controls once airborne out of Wright-Patterson, with Raeford's admonition to "Wake me when we're over the lake" on approach to Eglin.

So, he always flew the aircraft back, but when overhead Lake Jackson in Florala, Alabama, Ward would punch the pilot to wake him up and transfer control.

In the earlier photo, Ralph (left) is shown in his Air Force uniform with his brother, Julian, in his Army uniform. Julian had quite a war record of his own. He fought against the Japanese in Burma, on one extended mission hiking more than 400 miles through the jungles and mountains.

After the end of World War II, Julian aligned himself as an adviser with Generalissimo Chiang-Kai Shek of the Nationalist Chinese who were being attacked by the Communist under Chairman Mao Tse-tung. Chiang and the Nationalists removed themselves to Taiwan after losing mainland China to Mao.

Ralph had another brother who served in the Navy, Chief Petty Officer Lloyd Lester Ward, who ultimately reached grade of W-2 (warrant officer, grade 2). His combat record was outstanding as he earned 16 battle stars aboard the USS *Buchanan.*

At the end of the war in the Pacific, CPO Ward's ship delivered General Douglas MacArthur into Tokyo Bay so he could go aboard the USS *Missouri* to receive the Empire of Japan's surrender. After the ceremony, the *Buchanan* took MacArthur back to the port of Yokohama.

On another tour not involving flying, Ralph was stationed near the Arctic Circle, working as a radar operator. His job was to monitor Soviet bombers as they flew inbound missions up to America's border, testing the United States warning and defensive systems.

Ward's service there was in the early years of Strategic Air Command (SAC) at a time when the U.S. had bombers airborne around the clock in a display of deterrence against the Soviet Union.

During the Korean War era, Ward was a recruiter. He also worked at Eglin AFB where he hauled rocket fuel for missile launch preparations. He at one time was also a first sergeant as part of his many duties. He said he once was part of a group that met and hosted President Harry Truman at Eglin AFB, and he persuaded the president to eat lunch with several of the NCOs.

Furthermore, one of Ralph's sons made a career of the Navy. Captain Glenn Howard Ward had a wonderful career of 30 years in the Navy, then continued to serve his country as a civilian working for the Department of Defense in a position where he continues to serve today.

Ward recalls some of the music of his generation. Glenn Miller's *Chattanooga Choo Choo, In The Mood,* and *Over The Rainbow* were favorites. Tommy Dorsey's *Stardust* and *In The Still Of The Night* were other songs of his era that were "easy on the ears." [3]

In a family full of patriots, Ralph Ward served his country with distinction for 25 years, finishing his career as a senior master sergeant. Now retired and 89 years young as of this writing, he lives in Prattville, Alabama, and has so many fond memories of his experiences. A grateful nation thanks him and his brothers, as well as his son, for their longstanding service to our national defense.

CHARLES DURNING (ACTOR)

"June 6, 1944, D-Day...He was the lone survivor in his unit that day"

Born February 23, 1923 into an Irish family with ten children in the small town of Highland Falls, New York, near West Point, Charles Durning exemplified the "ordinary citizen who performed extraordinary service" to his country in its time of need. He always rose to the occasion, whatever the circumstances.

I never met Charles Durning, but I admired him from a distance for his service record and from having seen many shows with him portraying one character or another. With him being from a large family that often struggled during lean times (as did my own large family), there was always a feeling that I identified with and appreciated many of the things he endured as a youngster. He succeeded against big, big odds, as did many of those described herein.

Durning grew up in a difficult time with poverty all around because his dad had lost a leg and was gassed during World War I. With his father unable to work, his mother did laundry to put food on the table for her children.

During the June 6, 1944 D-Day invasion at Normandy, Durning was in an Army Ranger unit that was among the first to go ashore. He was the lone survivor in his unit that day.

On that fateful day he killed several German soldiers and was himself wounded in the leg and bayoneted by a German soldier before he killed the German with a rock he found laying near the beach.

Durning also fought in the Battle of the Bulge, but was captured and survived a massacre the Germans inflicted on their prisoners. Though his unit was eventually defeated in Belgium by an SS Panzer unit, Durning escaped and was spared the fate met by many of his friends—the infamous *Malmedy massacre*, in which a German officer, Joachim Peiper, had over 100 American prisoners shot dead without warning as they stood in a field.[1]

He got his start in show business as an usher at a burlesque theater in Buffalo, N.Y. when one of the comedians showed up too drunk to perform, and he was asked to fill in.

After the war, he became the most recognizable character actor in Hollywood, starring in numerous movies such as "The Best Little Whorehouse in Texas," "North Dallas Forty," and "O Brother Wherefore Art Thou."

In his 50-year acting career, he played roles as the Pope, Nazis, and once played a "suitor" of Dustin Hoffman in the movie "Tootsie." [2]

One of his most notable roles included playing a crooked cop stalking con men Robert Redford and Paul Newman in the movie "The Sting."

Durning's war-time decorations include the Silver Star, Bronze Star, and three Purple Hearts from injuries suffered.

He passed away in December 2012, at age 89, but not before leaving a legacy of valor and supreme patriotism as a soldier, and then winning the hearts of millions of Americans who watched him in so many award-winning acting performances.

RALPH, JOHN, AND TONY PEACOCK

Anthony "Tony" Peacock at his home Jan 12, 2013

An in-law I admired, Ralph Leslie Peacock, was born May 18, 1918, near Cairo, Georgia, to Leslie Hugh and Annie (Aldridge) Peacock. The family lived and grew up in the nearby Boles Springs community of Grady County. He graduated Cairo High School in 1935.

Ralph first married Clarice Carroll of Carrolton (who, with Ralph before Clarice's passing, raised a son Ralph, Jr. "Butch" and daughter Reesa into their teen years).

After Clarice's death, Ralph later married this author's sister, Hettie. It was during this time that I became acquainted with Ralph and came to admire him as a person and appreciate his service to the nation. He had a wonderful sense of humor. Butch and Reesa were not much different in age than myself and younger brother Tony when we were growing up.

One of three brothers to serve in uniform, Ralph, Sr. joined the Army during World War II. He was trained as a bombardier and armor handler, and received one Purple Heart. He was assigned to the 385th Bomb Group, 8th Air Force, flying out of Great Ashfield, England, in Suffolk.

Most B-17s had thirteen .50 caliber air-cooled machine guns for defense against enemy fighters, and every man on each 10-man aircrew had to know how to fire the guns, which spit out 550 rounds per minute. The 385th BG operated primarily as a strategic bombardment organization until the war ended, striking such targets as industrial areas, air bases, oil refineries, and communications centers in Germany, France, Poland, Belgium, the Netherlands and Norway. [1]

In discussing his service, Ralph told of how some of the younger aviators would nonchalantly declare before a combat flight that they didn't need their flak vest, or they wouldn't wear a flight jacket, or some other piece of gear.

Ralph would gather all the crew's gear up and sit on it in the early phase of the flights, and by the time the flight was getting to altitude or crossing into enemy territory, those same ones would be approaching him and yelling, "Hey Ralph, where's my stuff? I need it!"

B-17 bomber from WWII era (photo with author's grandson Josh), Eglin AFB, FL Armaments Museum

One noted author described some of those missions as:

"Losses among aerial gunners were exceptionally high in 1943. Aircrews were provided with parachutes, but nearly all refused to wear them—with macho pride. The chutes would pile up in the waist. When the bail-out bell sounded, men would grab their chutes—if they could find them. For gunners in ball and tail turrets, chances were they wouldn't have enough time to get out, reach the waist, snap on a chute, then leave the plane. A stricken bomber was likely to go into a spin very quickly, and its centrifugal force penned men to the walls." [2]

Ralph Peacock later cross trained into the Personnel field and ultimately served two tours in Vietnam, the last one for thirteen months at Cam Ranh Bay and Ton San Nhut, South Vietnam. At Cam Ranh Bay, there was a Viet Cong encampment only about two miles away, and his location was attacked and almost overrun. When he left Vietnam, he was decorated with a Bronze Star with "V" for Valor.

At retirement in 1971, Chief Master Sergeant Ralph Peacock, Sr. had served several years as Sergeant Major for the Armament Development

Test Center under the command of Major General Jewell Maxwell. By 1971, he had a total of 23 years service in uniform.

After his Air Force years, Ralph became quite the fisherman, as he perfected the techniques of catching Pampano, a delicacy. He had other hobbies, such as a greenhouse, and he really enjoyed going to the Southwest U.S., collecting raw stones, and turning them into polished stones of gem quality. He was also Treasurer of the Eglin Federal Credit Union, Shalimar, Florida, before he passed away 10 Aug 78.

Ralph had an older brother in uniform, John Robert Peacock, born in Grady County, Georgia, March 25, 1913, who was a Radio Technician in the Signal Corps. His service began with enlistment on 5 Feb '43 and training at Fort Meade, MD. He was in the 925th Signal Dept. Co., Aviation, part of 8th Air Force.

John and a good friend had married Whigham sisters (John's wife was Millie). During a furlough, John decided to visit that in-law, Mrs. A.A. Giddens, who was in London. On the return flight on 1 Nov '45, John's plane crashed, and he and all aboard (some 25) were killed.

During foggy conditions, the plane hit a mountain in the French Alps. Later investigation indicated there may have been an instrument error, most likely the altimeter. He's buried in Cairo, Georgia.[3] John was just another American who paid the ultimate sacrifice in service to his country.

Ralph had a third brother in uniform, Hugh Anthony "Tony" Peacock, now age 83 at time of this writing, with wife Helen (Willis) in Milton, Florida. Helen was from Thomasville, GA.

Tony was trained in Army Field Artillery and served in the Occupational Forces in Japan on Honshu Island, following Japan's surrender to end WWII. Tony arrived in country at Yokohama aboard a *Liberty Ship* as part of the 4th Replacement Depot. He didn't think too highly of Liberty Ships, saying it was the worst trip he ever made, as the ship "creaked noisily" every time it hit a wave crossing the Pacific.

He served in the 8th Field Artillery Battalion—manning 105-mm guns—which at the time was part of the 25th Division, 8th Army, I Corps. Along with the 24th Division, the 25th had been one of the first to be deployed by General MacArthur at the outbreak of hostilities in June, 1950, in Korea.

He was a "rod man" at the end of the gun barrel, setting elevations for the artillery firing. He was also in the Motor Pool driving a vehicle pulling artillery cannon. He later became a medic, mostly giving shots and performing clerical duties until he left the service in 1947.

In civilian life, Tony attended the University of Florida, getting BS and MS degrees in Agriculture. He and his family lived in Griffin, GA, where he worked as a cotton breeder for the USDA at the Georgia Experimental Station.

Later, when his son Ray was only seven weeks old, they moved to Ames, Iowa, for 3½ years where he completed a Doctorate in Agriculture at Iowa State University.

He later moved to Milton, Florida, where he ran the University of Florida Experimental Station in Santa Rosa County, experimenting with cotton, corn, soybeans, peanuts, and small grains. Tony worked for 41 years in that field, serving our nation in providing food products.

In addition to son, Ray, he and Helen had daughters Mary, Beth, and Linda. They lost Linda in a tragic accident while living in Griffin. [4]

The service of the Peacocks was typical of so many great families in World War II, as many brothers served simultaneously in an era of enormous patriotism and unified effort to win a war that ensured the survival of our culture and freedoms.

NBC Newsman Tom Brokaw referred to them as "Our Greatest Generation," and the Peacock brothers contributed much to that label. They certainly did their part in upholding the finest traditions of service to our country, not only in uniform, but in civilian life as well. The Peacocks are to be applauded for their longstanding service.

CHARLES P. "CHUCK" HUNDLEY

"The belligerent Russians harassed Hundley and other Allied pilots"

Chuck Hundley and author, 2012 Ft. Walton Beach, FL (personal photo)

Retired Air Force Lieutenant Colonel Charles P. "Chuck" Hundley was born March 17, 1922, in Dunnsville, Virginia. After high school, he attended Lynchburg College in Virginia until joining the Army after World War II broke out. At the time I met him, he was only a few months shy of 91 years old and a resident of Seminole, Florida (not far from St. Petersburg).

Hundley was a B-24 *Liberator* pilot in World War II who flew numerous high-risk bombing missions on the oil fields and refineries of Ploesti, Romania, in an effort to break the back of the Nazi war machine. Those oil fields by some estimates provided 60 percent of Germany's crude oil.

The first attack on Ploesti took place on 1 August 1943 and involved 177 B-24s with 1,726 crewmen who departed Benghazi, Libya, to make the attack on the fields north of Bucharest. Some 54 of the aircraft and 532 men never made it back. Five Medals of Honor were awarded for the daring attack. Of the aircraft that made it back, more than 50 had battle damage. Hundley was wounded on one of the missions and received the Purple Heart.

Resistance over the target was strong. There were 200 German 88-mm anti-aircraft guns, plus deadly 37-mm and 20-mm guns as well. More than 300 *Luftwaffe* and Romanian fighter aircraft came up to attack the B-24s.

Those of us who grew up watching war movies made by the Army in WWII have likely seen depictions of aircraft such as the one that another Ploesti raider—Lieutenant Lloyd Hughes—was piloting. He was leader of his formation who held heading after his aircraft was hit, dropped his bombs, turned away, and had a wing break off, causing his aircraft to immediately crash to the ground with all aboard killed. He was one of five to receive the Medal of Honor. [1] That could just as easily have been Chuck Hundley; he was just more fortunate than Hughes and many of the others who gave all.

The spring of 1944 saw more raids on Ploesti, and Romania surrendered on 24 August 1944.The next month, Hundley married his sweetheart, Vinette Wheeler, and they would be married for 59 years until her passing.

After the defeat of Germany, thousands of Soviet Union troops surrounded Berlin in June, 1948, and blocked the Western Allies' railway, road and canal access to the sectors of the city under Allied control.

Their aim was to force the western powers to allow the Soviet zone to start supplying West Berliners with food and fuel, giving the Soviets practical control over the entire city. In response, the Western Allies organized the Berlin Airlift to carry supplies to the people in West Berlin.

Later, Air Force Lt Gen William H. Tunner described the operation as, "Two and a half million people had to have supplies to survive – primarily food to eat, coal for warmth and light – and yet every surface route was closed to them. We moved into that situation with the Berlin Airlift. It lasted over thirteen months; in the last four months we were flying nine hundred round trips a day, delivering the necessities of life to the people of one of the world's largest cities." [2] Hundley said aircrews from the United States Air Force, the British Royal Air Force, the Royal Australian Air Force, Royal Canadian Air Force, Royal New Zealand Air Force, and South African Air Force flew over 200,000 flights in one year and provided more than 4,500 tons of daily necessities such as fuel and food to the Berliners. Chuck Hundley stated he and his crew flew 225 of those airlift missions.

The belligerent Russians harassed Hundley and the other Allied pilots through a variety of measures. At Gatow field, where the Russian Zone came right up to the end of the runway, one night the Russians moved powerful searchlights into position and began flashing them into the eyes of the pilots taking off. As time went on, they harassed in other ways. On occasion they staged anti-aircraft practice, with a plane towing a target for the guns below to shoot at. Sometimes the shells burst in the corridor the Allies were flying. They even had a Russian plane tow a target and had their fighter aircraft pour machine gun bullets into the target.[3] The intimidation tactics only made Hundley and the other pilots more determined to succeed with the airlift.

By the spring of 1949, the effort was clearly succeeding and, by April, the airlift was delivering more cargo than had previously been transported into the city by rail.

The success of the Berlin Airlift brought great embarrassment to the Soviets, who had refused to believe it could make a difference. The blockade was lifted in May 1949 and resulted in the creation of two separate German states, The Federal Republic of Germany (West Germany) and the German Democratic Republic (East Germany) split up Berlin.[4]

Chuck's long-time friend, retired USAF Major Ralph Ingles, of Alexandria, Louisiana, escorted him to a reunion of the Air Commando Association on October 13, 2012 in Ft. Walton Beach, FL, and snapped the included photo.

Hundley and Ingles were in the same flying outfit on more than one occasion, and Hundley was an instructor pilot for Ingles in C-119s (and other aircraft they flew) in the 39th Troop Carrier Squadron in Munich, 1955-58.

After he retired from the Air Force in 1970, Chuck and Vinette moved to St. Petersburg. They both entered real estate and worked together in business. They were also long time members of the St. Pete Yacht Club and the U. S. Power Squadron, of which Chuck was commander for two years.

Chuck and Vinette had daughters Beverley (husband, Bob Allen), and Charlotte (husband, Dean Young). Chuck passed away in Virginia on February 26, 2013.

Another of our Greatest Generation gone, but certainly not forgotten. I'm thankful I had the opportunity to meet and talk with someone like Chuck Hundley who made such a difference to the national security of our nation.

WILLIAM NELSON MURDOCK

"Billy went ashore Utah Beach on June 6, 1944..."

William Nelson (Billy) Murdock was born July 27, 1925, to Phillip M. and Beuton (Mayfield) Murdock in Kentucky. The Murdock's were farmers on land that at present has been in the family for six generations.

Billy graduated from Lynn Grove High School (south of Paducah, in Calloway County) in May, 1943. Shortly after his 18th birthday, he received a draft notice.

He left Murray, KY on a Greyhound bus with a group of draftees on October 8, 1943, heading for Fort Benjamin Harrison, IN for processing.

Billy had taken Aeronautics in high school and thought that he would like to be in the Army Air Corps. He passed all written tests to go to training to be a second lieutenant in the Army Air Corps, but his hopes were dashed when he learned he was color blind.

April, 1943, he was inducted into the Infantry and sent to Camp Wheeler, GA near Macon. After basic training he was given a 2-week furlough with papers directing him to report back to Fort Meade, MD in April of 1944. After a short stay there, he boarded the luxury liner *The Queen Mary* for a "cruise" to England.

The ship had been stripped of all her grandeur, painted gray, and filled with Army Barracks. The grey was so that she would be hard to spot in the foggy Atlantic Ocean by Hitler's U-boats that were determined to stop the ship from transporting American soldiers to England. The ship's officers were given strict orders to go as fast as they could and not stop for anything. The food was mostly Sauerkraut and hard tacks, and Billy didn't care for the "cuisine" aboard his cruise.

Sometime in early April, Billy landed in England and the soldiers camped in a sheep pasture where they continued training for the D-Day Normandy Invasion. He was with Company C., 12th Infantry Regiment, 4th Infantry Division of the U.S. Army.

Billy went ashore Utah Beach on June 6, 1944, and lost many friends and his best buddy almost immediately. He was first wounded in the leg, but it was not too serious. The next was worse.

Billy was a designated Scout for his company. It was his job to slip into the combat zone, locate the German troops, and report back to his CO. On one occasion, as he reported back to this CO, a German sniper shot his commander through the head, and the bullet continued through Billy's shoulder, causing a much more serious injury.

The Infantry was very much involved in heavy combat and did not receive any mail until August 2nd. Thank goodness, Billy received a large bundle of mail that day because, on August 4, 1944, when returning from a scouting trip, the Germans hit him in the back with a mortar shell, sending shrapnel through his lung, into his spine and many parts of his body.

He remained unconscious and in a field aid station for several days. The critically wounded were left there with a doctor giving them the needed medication until they died, or, if they showed signs of living, they were then sent to a hospital. After several days, when he regained consciousness, his doctor said, "Son, you are going to make it," and he was transported in a plane to a hospital in England.

His family did not hear from him for a long time, but finally his mother received a telegram from the War Department that said, "We regret to inform you that your son, PFC William N. Murdock, has been critically wounded in France. We will contact you when we receive further information."

Billy thought he would be sent back home when he was able to travel, but that didn't work out as he surmised.

In December, 1944, Billy was still convalescing in England when he was moved back to France to work in a Post Office to relieve another soldier to help liberate the American soldiers that were surrounded in the Ardennes Forrest area in Belgium.

Hitler had surrounded the American and Allied troops with three Armies of at least 20 divisions. They had failed to close the gap at Bastogne, and GEN George Patton and GEN Omar Bradley moved 3rd Army Infantry reinforcements in so that the Allies were able to strike the Germans from the middle of the bulge. They soon beat the Germans back. This battle went down in history as "The Battle of the Bulge."

The winter was extremely cold and harsh. The soldiers suffered terribly, many had severe frost bite, and many died. They did not have warm places. Billy suffered especially, because his lung was still weak from his wound. In April, 1945, his family learned he was still in France, but as soon as the war ended, he was sent into Germany. He and others slept in old train cars and burned anything they could find, including old tires, for warmth. Their faces became black with soot, but staying warm was critical.

The Army had a point system on who could be released, with 60 needed. Billy had 59 points. On July 6, 1945, he was allowed to write his first *Blue Envelope* letter to his mother and girlfriend, Jean.

Jean later recalled that he could not tell them any specifics, but it became apparent he was headed to another war zone. According to Jean, "The letters were all strictly censored. One of his friends sent me a letter and drew a canoe with a soldier in it on the front of the envelope. I was supposed to figure out that he was on the water headed to some far port, and that was not a home port." [1]

He had been transferred to the 731st Engineer Depot, Co. A, APO San Francisco, CA. He was sent through the Panama Canal en-route to Hollandia, New Guinea. His luck was beginning to change—a little. On August 15, 1945, when they arrived they could hear loud horns blowing and the Navy sending up red, white and blue flares. An armistice was about to be signed there.

They stayed on the ship, the USS *S. D. Sturgis* and went from there to the Philippine Islands. September, 1945, they landed on Manila, and on Sept. 6, 1945, he sent home a letter from San Jose, Luzon. Next, he sent a letter from San Fernando Island, saying they were leaving that location.

Billy went on to Kuril Island, north of Japan via an L.S.T. These boats had a flat bottom. They went through a terrible tropical storm, and it really tossed them around. These Northern Islands were the string that connected to Alaska. Their papers (orders) were to supply a construction battalion building and an air strip. Billy was in charge of the depot where building supplies were distributed. [2]

Letters home showed he was still at sea from October 10th through October 24th. He also said on November 20th, 1945, he and his buddies walked through the ashes of Hiroshima, Japan, just 14 weeks after the atomic blast that killed over 80,000. They likely didn't know the long-term dangers of radiation. His last letter home said on December 17, 1945, he was being sent to Nagoya, Japan to be shipped to California.

From California, he went by train to Fort Knox, Kentucky where he was discharged from the Army on Jan. 14, 1946. Heading south, he stopped in Paducah at 4 a.m. to see a young lady, Rebecca Jean Adams, a student at Draughon Business College, that he had not seen for over two years since he had said goodbye and boarded that Greyhound bus to head off to war.

Jean had waited those two years. According to Jean, "On February 23, 1946, he took that young lady home with him. We have had almost 58 wonderful years of happiness. I have loved him ever since I was 15 years old, in 1943, when he sold one of his bull calves, bought me a diamond ring and asked me to wait for him. I said that I would. I did, and I would do it again." [3]

Billy's decorations included two Purple Hearts, Bronze Star, European-African-Middle Eastern Campaign Medal with 4 bronze battle stars on the ribbon bar, Good Conduct Medal, Army of Occupation Medal, WWII Victory Medal, Asiatic Campaign Medal, Marksman Medal, Rifleman Medal.

He and Jean settled back in Lynn Grove and resumed farming like the previous five generations, growing tobacco, corn and soybeans. Billy beat back cancer that was discovered around 1958, possibly from that tromp through the ashes of Hiroshima. He'll celebrate his 88th birthday in a few weeks. He and Jean have three children, Carolyn being the oldest. Then, William Nelson Murdock, Jr. and James Ray Murdock, the youngest.

Billy's brother, Phillip Howard Murdock, also served in the military.[4] The Murdock family is part of that group of *just ordinary Americans performing extraordinary service for their country.*

ROBERT C. BENNETT

"Bennett finished his pilot training and flew his P-51s"

Robert Bennett official photo (Army photo 1940s)

Lieutenant Colonel (Ret.) Robert C. Bennett was born to Clarence and Jessie Ann (Jones) Bennett September 26, 1923, near Mount Olive Baptist Church, about five miles from McKenzie, Alabama. Clarence was a native of McKenzie, and Jessie was from Montgomery.

As a youngster, Robert was an acquaintance of Hank Williams, Sr., whom he referred to during my interview with him as, "The greatest country music singer of all time." He had attended several of Hank's performances in his youth, and he still remembers well the first time Hank was on the Grand Ole Opry radio show singing his *Lovesick Blues*. Like most all American servicemen, those of Bennett's era and from his part of the country were influenced by the music that was popular at the time. Most of them knew all the words to Hank's songs such as *Your Cheatin' Heart*, his famous *I'm So Lonesome I Could Cry*, and *Cold, Cold Heart*.

One would expect such from a young man who grew up within a few miles of Hank's birthplace near Georgiana.

Robert was drafted into the Army Air Corps in 1943, completed his basic training in Texas, then was trained to be an airborne gunner in advanced training. His first assignment was as a gunner aboard B-17 bombers based in England. From there, they bombed targets in several areas of Germany, including industrial facilities, Munich, and other cities.

After a combat sortie one day, his pilot took him aside and said, "Robert, you're just too smart to always be a gunner; I'm recommending you to become a pilot."

As a result, he attended flight training near Arlington, Texas (Dallas-Ft. Worth area), with a portion of training completed near Houston, including a few flights over the Gulf of Mexico. Some of the training was in the venerable T-6.

On the day he finished and was awarded his aviation wings, he also was awarded an officer commission and pinned on rank as a second lieutenant.

Bennett finished high enough in his class to qualify for fighter aircraft, getting an assignment to fly the P-51 *Mustang*, at that time the United States' newest fighter aircraft that was fast, nimble, and powerful with its 16-cylinder Merlin engine. And, it carried lots of gas.

The model Bs and Cs were capable of speeds up to 441 mph at an altitude of 29,800 feet, finally giving our vulnerable WWII bombers the long-range escort so desperately needed.

The D model included a new bubble canopy for better visibility, plus an increase from 4 to 6 the number of .50 caliber machine guns; 5-inch rockets and 2,000 pound bombs were also standard armament on some models. [1]

Bomber escort and air-to-air sorties were critical in WWII. Though the war in Europe was over by the time Bennett finished his pilot training and flew his P-51s from an airfield near Paris, France, Bennett trained hard on what was called "air-to-air" in case hostilities broke out again with other potential rivals (such as Russia and other Eastern Bloc nations). After the war in Europe had ended, he flew numerous patrols along the line dividing American and Russian occupation forces in Germany, "just to make sure the Russians followed the agreed-upon rules."

On one memorable sortie, shortly after taking off and during his climb out, his P-51 engine failed completely. Fortunately, he had enough altitude that he could glide back to the airfield and complete a "dead-stick" landing. He flew the P-51 for a total of three years.

While assigned to Page Field at Fort Myers, Florida, Robert married Dorothea Cook (of McKenzie) on May 5, 1945. Her parents were Lewis and Grace (Parker) Cook, also of McKenzie, and one of her brothers was Hampton "Hamp" Cook, who for a few years was the head football coach of my hometown high school team in Opp where he turned out some outstanding teams.

Robert and Dorothea lived in base housing, and Robert remembered she absolutely loved everything about their being in the military.

On one assignment, she accompanied him to Europe where they were assigned to a base near Munich where the base runways ran north/south, with the north end of the runway being only a short distance from a window in their apartments.

On occasion, when he knew she was near the window during his departure, he would hold the aircraft on the runway as long and as low and fast as possible so as to "buzz" her on a close flyby, something she never seemed to appreciate!

The Air Force retired the P-51s when jet fighter aircraft were ushered in, and Robert was subsequently assigned to fly C-47 cargo aircraft in the post-war period. In his career, he flew the C-47 "Gooney Bird" more than any other aircraft.

His next flying assignment was at Lockbourne AFB in Columbus, Ohio, flying the B-47 Bomber, which was flown with "hard crews" assigned. His co-pilot was John Lally, and his navigator was Riley Tieg. For several years he flew the aircraft, including overseas deployments, but never dropped a bomb in any combat action.

The B-47 was a swept-wing, jet-turbine-powered bomber that had a length of 107 feet and a wingspan of 116 feet. It was powered by six GE J-47 engines, each capable of generating 5,200 pounds of thrust.

The aircraft carried 10,196 gallons of fuel in fuselage tanks. It weighed 79,000 pounds empty, carried a maximum bomb load of 20,000 pounds, and could cruise at a max cruise speed of 448 knots at a maximum of 41,000 feet altitude, with a combat radius of 2,000 miles.[2]

Per Bennett, it was the most modern and fastest combat aircraft in America's inventory when first introduced.

After flying the B-47, Bennett went back to flying C-47s. He said he was his squadron's Assistant Operations Officer, and achieved the rank of lieutenant colonel, which was the rank at which he retired on March 31, 1965.

After retiring from the Air Force, he and Dorothea moved to Pensacola, Florida, where he worked for eight years as office manager for *Snelling and Snelling Placement Services*. Dorothea was his secretary, and they thoroughly enjoyed their time on the coast.

He was very good at his job, so good that he once won a new Cadillac in competition that named him "Manager of the Year."

A civic-minded individual, Robert was a leader in the Lions Club in Pensacola, where he established a nature trail at the University of West Florida, earning him accolades as "Lion of the Year." He was also a Shriner in an organization that raised money for burn victims.

At the same time, he managed to still do some flying in small single-engine aircraft. He had a flight instructor's certificate and trained several young pilots up to their solo flight and/or until they got their private pilot's license.

When it came time to completely retire, he moved the family back to McKenzie and built a new home. Dorothea passed away eighteen years prior to my interview with Robert (interview was on January 22, 2013, at which time he was only eight months shy of his 90th birthday). [3]

I might add that his handshake was still very strong, and Bennett got a twinkle in his eye while talking about his flying days, his time in the Air Force, and his life with Dorothea and their three daughters, Barbara (Jack) Stein, Cindy (Don) Leaders, and Cathy, who had passed away some time prior.

Robert Bennett's story is truly amazing. He was just a young kid from a town of barely 500 citizens, but when his country came calling, he answered the call, and he answered it in fine fashion. As a result, he had some of the greatest adventures one can imagine.

Robert Bennett passed away on June 17, 2013, leaving behind a legacy of patriotic service to his country and love for his family and fellow man.

It's not very often that a young man from his circumstances comes to fly some of our nation's highest performance aircraft and be part of the historic events with which he was associated. His was an amazing story, and one that adventurous youngsters of today might want to emulate.

HUBERT T. SULLIVAN

"The Roer River line, thirty-five kilometers long, burst into a ball of fire"

Hubert Sullivan (Army Photo)

Hubert T. Sullivan was born in Castleberry, Alabama, on February 14, 1924, to parents Joseph C. and Essie (Mixon) Sullivan. His father ran a sawmill and gristmill in that area, and Sullivan and his siblings attended Conecuh County High School (now a junior high school).

Hubert was drafted into the Army in 1943. He also had two older brothers—Joseph M. and Bert W.—who served in the Army Air Corps during World War II. Both were in Communications, with the older serving in Panama, and the younger in Alaska. A younger brother—Cecil D.—served as a flight engineer aboard tanker aircraft in the U.S. Air Force, after the end of WWII.

Hubert was first sent to Camp Blanding, Florida, for training, then shipped via train to Camp McCoy, Wisconsin, arriving there on October 7, 1943 in heavy snow conditions.

His Basic Training was followed by advanced training as a Combat Engineer. From Camp McCoy, he was sent to Camp Kilmer, New Jersey, for processing to ship out for Europe.

Sullivan and hundreds of others sailed out of New York Harbor for Liverpool, England. The ship was the HMS *Large Bay*, a British ship that had been used in World War I to transport cattle, so the poor shape that ship was in can only be imagined.

He and his company had boarded the ship a week prior to sailing so they could be checked out on manning the defensive weapons onboard. The guns included what the British referred to as "a 12-pounder." There were also 40-mm *Bofors* cannon for self defense.

The trip to Liverpool took 13 days, and happily there were no German U-boat attacks along the way. The *Large Bay* was part of a huge convoy. Sullivan said he "could spot convoy ships fore and aft as far as the eye could see."

Upon arrival in Liverpool, everything was blacked out because of the constant threat of air attack from the German Luftwaffe. After unloading, they were shipped via train toward Bournemouth on the southern coast of England.

An odd thing happened when his car was cut loose at a junction, so they ended up in South Hampton which required they be put aboard a private civilian train to head back to Bournemouth, their original destination. They later went back to South Hampton from where they sailed for Normandy, France.

The first wave of Combat Engineers with Sullivan went ashore Normandy in August, 1944 (after D-Day on June 6th), with a mission to clean up unexploded mines left on the beaches . He was assigned to the 9th Army, 251st Combat Engineer Battalion, under the leadership of American General Simpson, and attached to the command of British General Bernard "Monty" Montgomery.

In addition to clearing mines, his unit also built bridges, both pontoon and those of standard construction. They only built one of pilings and girders, as it sagged in the middle from the weight of tanks and other heavy equipment crossing. Lesson learned!

They fought their way across France and the Netherlands. In Holland, with so many low-lying areas and lots of standing water, the roads gave way easily because the road bases were often fragile and required a great deal of repair.

On some occasions the Army demolished a brick or cement structure and used the debris to shore up the roads for crossing.

Their route took them toward Aachen, Germany, as the Allies pushed the Germans back toward Berlin. His unit was considered an "illegitimate unit" and was called on for just about any type action.

As they approached the Siegfried Line, one call came for someone to use a flame thrower. Another soldier in his unit used the weapon to take out a machine gun "pill box," so Hubert was a little relieved not to have to do that job.

A huge natural barrier was the Roer River, site of tremendous battles along the Siegfried Line, where his unit crossed between Lynch and Roers on February 23, 1945. They crossed under a big artillery barrage from the Allies.

One distinguished author described what happened, as follows, at the time Sullivan and his fellow combat engineers approached the river.

At 0245, February 23, the Roer River line, thirty-five kilometers long, burst into a ball of fire. It was one of the heaviest barrages of the war. Big guns, ranging from 75 mm to 240 mm; heavy and light mortars; rockets; direct fire from tanks; .50-caliber machine guns and 40-mm anti-aircraft guns—every weapon the Americans had hurled against the enemy a forty-five-minute deluge of bullets and high explosives designed to stun, kill, or drive him from his position.

Men who were there were men who had been through many; their testimony is that this was the biggest one they ever saw. Ninth Army alone had more than 2,000 artillery pieces firing 46,000 tons of ammunition. [1]

Sullivan remembered that, "The British tanks of General Montgomery had no rubber shoes on the tracks as had the American tanks, so their tanks tore up the approaches to the bridges pretty badly and slowed the crossings."

When they got to the Rhine River, they followed pretty much the same pattern until they approached the Elbe River, which was the location where Germany's generals eventually decided to surrender and end the war in Europe.

Sullivan's unit crossed the Rhine using plywood M-2 boats. On some patrols, in an effort to help get German Prisoners of War (PW)'s across so they could surrender, Americans would wait until near dark to use their binoculars to check for gun emplacements so the Germans still resisting could not see the glint of the lens.

On one occasion, he remembered waiting by the river bank for an extended period for an expected PW to cross. Hubert and others placed two .50-caliber machine guns on the banks along the Allied side of the river. The PW never showed up.

He and others were always leery of friendly fire, as well as that of the enemy. To minimize that possibility, a system of challenge and response code words was used, his unit's challenge being "3" with the correct response being "5."

On one occasion, they loaded on a boat and headed downstream, drifting under an outpost of the 102d Infantry; he and his buddies were just as worried about fire from the 102d friendlies as they were about fire from the Germans.

Sullivan remembered the Germans' machine guns included "34s" and "42s" (the 42s could spit out the lead at a horrendous rate).

Hubert himself carried a Thompson sub-machine gun (.45 caliber)—not one with the drum cartridge, but one with sleeves.

At the Elbe River the Germans gave up. Adolf Hitler shot himself on April 30, which was big news!

Sullivan also recalled the time when word arrived that Franklin Roosevelt had passed away. One of the motor pool sergeants climbed a power pole of a communications line that was still active and tied on to verify the message. The news was all over the BBC radio network as well.

His unit had planned to cross the Elbe, and Sullivan had dreaded the thought, but when the Germans surrendered, he and others were moved back to France to set up a tent city referred to as *Camp Lucky Strike* near Rheims.

It was there that American soldiers were evaluated and processed to either return to the States, or head for Japan to join the fighting in the Pacific Theater.

His unit used PW labor to erect the tents. He recalled many German PWs were very emaciated from their having been on the front lines, fighting with limited rations. Germany's economy had collapsed, and supporting its troops became harder in the waning days of the fighting.

When American troops cleaned their plates at the mess hall by scraping residual food into the garbage cans, the PWs would often scrounge the garbage and grab a handful of discarded peas or beans, or even coffee grounds, and "gulp them down like a starving dog" (in his words).

After the Germans' surrender, Sullivan did not have enough service points to be discharged, so he attended one quarter of "Electronics" school in London at their Northern Polytechnic Institute. Afterward, he went back to France with the 371st Construction Company, where they did very little.

He did make a one-week trip to Switzerland on an R&R where he managed to get some pictures with some pretty girls, every sailor and soldier's dream in that day and age.

He sailed on an American ship back to New York Harbor, arriving with a bad case of tonsillitis. He asked to just have his throat swabbed with *Merthiolate* so they wouldn't keep him from his discharge. He did spend one week in the Fort McPherson Hospital near Atlanta before riding the bus to Montgomery where he spent the night with his dad's sister.

After the war, on August 3, 1947, Sullivan married Evelyn Pate, his teenage sweetheart from his Castleberry days.

He went back to school and completed a Bachelor's Degree in Education at the University of Alabama in 1949, then completed a Master's Degree there in 1950.

He taught in the Anniston, Alabama public schools and was Assistant Principal of Anniston Junior High before relocating to Opp, Alabama to become Opp High School principal in 1954.

He held that position until 1959 when Education Board Chairman Fred Lyons convinced him to go to work with the Opp and Micolas Cotton Mills as head of Personnel, Human Relations and Industrial Relations. He was with the mills for more than 32 years.

Sullivan was also very active with numerous civic organizations such as the Rotary, American Cancer Society, Alabama Chamber of Commerce, and Opp-Micolas Educational Foundation (which provided student scholarships, something of which he is very proud), and others.

In another recollection, Hubert remembered a former German PW coming to America through Canada, marrying a girl from Florida, and ultimately moving to Opp.

His name was Hans Werner, and he lived in Fred Lyons' former house. On an occasion at a social event, when discussing the war, Hans stated that he too had plundered the Americans' garbage cans for scraps of food. That is something not likely to be forgotten by any soldier who has been hungry and cold on the battlefield under wartime conditions.

Hubert's first wife Evelyn passed away in 1986. He married present wife Rebecca (Sullivan) some 18 months later. Rebecca is from the C. J. Sullivan family that owned and operated Sullivan Furniture Co. for many years, with stores in Opp and Florala.

In recognition of his lifetime of service to his country and his adopted town, Sullivan was inducted into the Opp Chamber of Commerce Hall of Fame in 2011.

He and Rebecca presently enjoy the retired life, attending church, and getting visits from their many friends in the Opp area.[2]

JAMES FORREST "FROSTY" COOK

"It would have blown sky-high, with nothing but a greasy spot left."

James F. "Frosty" Cook 1943 (Navy photo)

James Forrest "Frosty" Cook and his twin brother, John Horace Cook, were born as the youngest children to Grover Cleveland and Laura (Phelps) Cook on January 10, 1923. They spent their very early years in the Pine Level community in Pike County.

The family moved to the Opp, Alabama area where his dad farmed at a place on Pittman Road (near the present waste treatment lagoon).

Frosty attended Hickory Grove grade school, which was along U.S. Highway 84, about three miles west of Opp. Cook walked to school each day, and he recalled his grades 4-6 teacher was Murray Brogden.

He later attended Opp City Schools, again walking from home to attend classes. Prior to his military service, he worked in the cotton mills for three and a half years, plus another six months in a mill near Opelika, Alabama.

On January 20, 1943 he enlisted in the U.S. Navy in the middle of World War II and was sent to San Diego for basic training. His twin brother enlisted in the Navy and went to Great Lakes Training Center near Chicago.

Another brother, Roy C. Cook, joined the Army, and two other brothers, Grover and Kermit, also joined the Navy, an amazing thing—five brothers all in the military—but also something not all that uncommon during the World War II era! Those who lived through that era often stated that a person could walk the city streets and see very few young men. They were all off at war.

Frosty's duty was as an Armed Guard on merchant ships (Liberty Ships) and included duty on the SS *George C. Yount*, the SS *William H. Dall*, the SS *Cape Charles*, and the SS *William C. C. Claiborne*.

The SS *George C. Yount* was outfitted with two 20-mm machine guns on the bow, two between the bridge and the stern, and four on the bridge, which was fairly typical configuration for merchant ships subject to enemy attack.

Other guns fired were a 5-inch/.38 caliber and 3-inch/.50 caliber shell (much larger ammunition than the 20mm).

Cook stated he fired at attacking Japanese planes, but he did not take one down, to his knowledge.

The SS *Yount* was 441 feet long, had a 28-foot draft, and a speed of 12.5 knots.

Some of the more significant actions Cook saw were at the battle of Okinawa where he witnessed the *USS New Mexico* battleship attacked on 12 May 45 by a kamikaze, with a resultant loss of 54 killed and 119 wounded. Frosty and a fellow sailor were on deck of the *Yount* tossing a football back and forth when they spotted the incoming attackers, then ran to man their weapons.

The attack on the *New Mexico* was the second kamikaze attack that caused a lot of casualties. A few months earlier (January 6, 1945) she was hit during the battle at Luzon, P.I., resulting in the loss of the *New Mexico's* ship's Captain Robert Fleming, as well as a British Brigadier General aboard who was Winston Churchill's personal representative to General of the Army Douglas MacArthur. [1]

Cook's ship also on one occasion pulled into Guadalcanal where there had been heavy fighting, carrying a load of timbers intended for use in constructing a dock.

On one trip they had thousands of 55-gallon barrels of aviation fuel aboard. In his words, if the ship been hit, "It would have blown sky-high, with nothing but a greasy spot left."

He spent many, many weeks at sea, but on one especially long trip, Cook boarded a train in San Francisco and rode to Portland, Oregon. The crew then boarded the ship, sailed to Vancouver, British Columbia, took on a load of wheat, then set sail south to cross through the Panama Canal.

They sailed up the east coast past New York and all the way to Palermo, Sicily, where they offloaded their cargo, then sailed back to New York. These trips took weeks and weeks.

After his two years and ten months of WWII service, Frosty farmed with his brother Roy for a while and married Clara Blair from near the Perry's Store community. They have one son who lives in Lakeland, Florida. Having celebrated his 91st birthday at the time of this writing, he and Clara live on 2nd Street in Opp, Alabama.[2]

MORGAN M. MOSELEY

"He spotted the Indianapolis due to a quirk in visibility and moonlight"

Ship's Cook 1st Class Morgan M. Moseley, U. S. Navy photo

Ship's Cook Specialist 1st Class Morgan M. Moseley was born November 16, 1921, to Homer and Ida (Whigham) Moseley in the small rural community of Letohatchee, Alabama, a few miles south of Montgomery. The family moved south to the Lowery community later, and he was in a family of eight children being raised alone by their mother through the days of the Great Depression. As a young man, he worked on area farms, including plowing mules before he enlisted in the U. S. Navy after hearing about Pearl Harbor.

Assigned in 1943 aboard the Navy cruiser USS *Indianapolis*, Moseley was in charge of the butcher shop, deep freeze area, and cold storage. The cruiser was Admiral Raymond Spruance's flagship and had 8-inch and 5-inch guns, plus 40-mm cannon and 20-mm *Oerlikon* guns, along with three small spotter planes.

The ship and crew had suffered the loss of nine sailors in a March 31, 1945, kamikaze attack during the invasion of Okinawa. The full crew complement was 1,196 sailors.

The *Indianapolis* has a special place in American naval history, suffering the greatest loss of crew of any naval vessel in American history. Following repairs after Okinawa, the ship and crew sailed July 15, 1945, from Mare Island near San Francisco. At Hunter's Point naval shipyard, it picked up components of the atomic bomb *Little Boy* to transport to Tinian Island where the bomb was to be loaded aboard the B-29 *Enola Gay* for the drop on Hiroshima, Japan, which ultimately occurred on August 6, 1945, killing some 80,000.

After dropping off the bomb components at Tinian, the *Indianapolis* sailed to Guam to change out some crew and pick up 250 new recruits, then sailed from Guam toward Leyte, Philippines.

In the wee hours of the morning of July 30, 1945 (two minutes after midnight), between Guam and Leyte at coordinates 134^0 E. and 12^0 N., the cruiser was struck by torpedoes from the Japanese submarine *I-58* which was commanded by Japanese Navy Lieutenant Commander Mochitsura Hashimoto.[1]

Some 60 feet of the bow was blown off the cruiser, with the loss of more than 300 from the explosion or drowning as they sank with the ship which went down in just 12 minutes.

Morgan Moseley was one of some 880 aboard who went into the water alive. Many were burned, bleeding, in agony from scorched lungs, broken bones, cracked skulls and concussions. Only about half the men had a life jacket, either kapok-filled vests, or inflatable rubber belts. Moseley was one who had neither.

He said the first to die were usually the burn victims. Others died from hypothermia when their body temperature dropped, or from exhaustion that prevented them from holding their head above water, or in many cases shark attacks. Moseley survived for five days and four nights by hanging onto a line attached to bigger life rafts that were bunched together with the most severely wounded in the rafts. Some sailors on the bigger rafts swapped out with Moseley when he became so exhausted he couldn't hold his head above water.

Moseley recounted that many of the sailors were covered by oil from the explosion. Many could not close their eyelids because of the oil. One blessing of the oil was that it helped to hold down the severity of sunburn for some sailors.

Drawn in by the scent of blood or urine, sharks began arriving from all directions by the second day the sailors were in the water. Moseley recounted instances where a shark would grab a sailor and pull him under. Then the sailor would surface again in a pool of blood, but missing a limb and screaming in horror. The shark would take him down a second time, and he would not be seen again.

With no fresh water to drink, thirst became a horrible problem. Many began to have terrible hallucinations, and some attacked each other with survival knives they had on their person. Some sailors succumbed to the temptation to drink salt water, despite warnings from a Navy doctor not to do so. Salt water within a few hours killed them when their kidneys could not get rid of body toxins.

By resisting the temptation to swallow salt water and the grace of God, Moseley was one of only 317 sailors who somehow survived. He felt sharks swim by and some scraped against him, but he was not attacked. Without fresh water, Moseley said he lost 45 pounds of body weight in his ordeal, dropping from his normal 205 pounds to 160 by the time of his rescue.

The *Indianapolis* was not reported overdue, so rescue efforts did not start for several days, and then only after a Navy pilot, Lieutenant Wilbur "Chuck" Gwinn, by chance spotted the oil slick and men in the water when he overflew the wreckage site in his Lockheed *Ventura PV-1* bomber during a routine patrol mission. Gwinn radioed for additional aircraft and rescue ships which began arriving quickly.

Morgan Moseley was picked up on the fifth day by the USS *Cecil J. Doyle,* a destroyer escort. He later was discharged from the Navy on November 12, 1945. Not surprisingly, he declined to re-enlist, despite promises of a big promotion.

Captain Charles Butler McVay, Jr., the skipper of the *Indianapolis,* was court-martialed after the war, facing Navy charges of not taking all precautions for the welfare of his crew. He faced two charges: not altering course—zigzagging, in Navy parlance—that some believed may have prevented the sinking; and failing to issue and insure the execution of timely orders for the abandonment of the ship.

Hashimoto testified at McVay's trial that he spotted the *Indianapolis* due to a quirk in visibility and moonlight that silhouetted the ship on the horizon for a short interval of time near midnight. Hashimoto said zigzagging would have made no difference in preventing a successful attack. McVay was convicted on the failure to zigzag, but acquitted on the failure to issue and insure the execution of timely orders for ship abandonment.

Receiving lots of hate mail over the years from deceased sailors' families and carrying a lot of guilt, McVay committed suicide on November 6, 1968, using his .38 caliber service revolver. Due to research efforts of Hunter Scott, a young school boy from Pensacola, the U.S. Congress in 2001 passed resolutions exonerating Captain McVay, saying in essence he was wrongly convicted in the court martial.

Modern day movie fans may remember the boat operator Quint in Steven Spielberg's movie *Jaws* who was determined to hunt down the great white shark that was terrorizing Amity Beach-goers on a holiday weekend.

As part of the script, after having a few drinks, Quint narrated how he had been aboard the *Indianapolis* when it sank, but he had survived the ordeal, much like Morgan Moseley survived, only Moseley's ordeal was real, not just something in a movie.

After the war, Moseley married Joy Boutwell of Jay, Florida. They had daughters Janna, Zanda, Landa, and Tanya. Moseley worked for the CSX and L & N Railroad for 38 years until retirement. He passed away in DeFuniak Springs, Florida, on February 25, 2013.[2]

Author with Morgan Moseley (October 7, 2012)

WILSON MERRILL SAULS

Merrill Sauls holds a map with a circle near Wesel, Germany where he parachuted into combat in 1945

Wilson Merrill Sauls was born in Rose Hill, Alabama, June 24, 1919, to parents W. A. "Bud" and Susan (Adams) Sauls.

His parents were farmers who had come to Rose Hill from near the China Grove Community north of Troy, Alabama. As a youngster, Merrill was very familiar with picking cotton in hot/humid fields, stripping cane and incurring lacerations of the hands, and, worst of all, the extreme unpleasant stinging of gathering velvet beans. He also did various and sundry other chores that go along with being a farmer's son.

Merrill dropped out of school in the 6th grade, did various odd jobs, then joined the National Guard unit, 1st Battalion-117th Field Artillery, in Andalusia in 1939 (the unit was part of the 31st Dixie Division).

He later would go back to school and get his diploma, but adventure awaited him first. His specialty in the Guard was "wireman" where he rolled out long lengths of spooled wire that were used for communication systems.

In 1940, his unit mobilized as Regulars, then he headed off to Camp Blanding near Jacksonville, Florida, for training that lasted nearly a year. Merrill recalled the soldiers were on maneuvers a lot in Mississippi and Louisiana after the Pearl Harbor attack on December 7, 1941, ("a day that will live in infamy," according to President Franklin Roosevelt).

Wanting something more exciting than lineman, he volunteered for paratrooper duty in 1943, not knowing that there was a minimum weight limit. When he reported for his physical exam, he was told he was three pounds too light, so they told him to come back the next day after eating a lot of bananas and milk, with a promise they would weigh him first thing in the morning. He did so and barely made the weight limit.

There was a reason for the minimum weight limit. A soldier that was too light could get caught in the slipstream on exiting the aircraft during jumps and might be severely injured when slammed against the fuselage by the wind.

His training at Fort Benning lasted some 4-5 weeks, then he trained at Camp McCall, N.C., where he "trained, trained, trained, and trained some more."

When he was sent down to the base in Savannah, Georgia (Fort Stewart), he trained as a machine-gunner on the .50 caliber weapon. A lot of the training consisted of shooting at a target towed by an aircraft over the Atlantic Ocean, then back for more paratroop training.

Sauls first went overseas on a brand new troop transport and arrived at Laharve, France, in 1944. There, they opened up crates of new weapons, cleaned the *cosmoline* (protective grease layer) off the new equipment, and prepared for combat.

He stayed in a unit, the 464th Parachute Field Artillery Battalion (for Operation Varsity's "Rhine Jump"), that was kept separate until combat. When they went into combat, they were attached to the 17th Airborne Division.

He first entered combat March, 1945. They got up at 0200 hours, ate breakfast, and were briefed by the colonel, a man Merrill described as "one of the finest, bravest soldiers I had ever seen."

The colonel stood erect and began "You men have been trained well. I won't ask you to do something I won't do. I'll be the first to go out when we cross the Rhine River. When I jump, you follow me."

Merrill said theirs was the second biggest invasion to date in the war, second only to D-Day's operation. They dropped from an altitude of 450 feet (50 feet below the normal minimum) because planners didn't want our troops hanging in the air too long and be easy targets for enemy gunners.

This was a time when General Patton was having trouble getting his tanks across the Rhine (Patton's Third Army crossed on March 22-23 at Oppenheim, near Mainz).

The mystique of Patton grew when, "The following day Patton walked across a pontoon bridge built by his engineers. He stopped in the middle. While every GI in the immediate area who had a camera took a picture, he urinated into the Rhine—a long, high, steady stream. As he buttoned up, Patton said, 'I've waited a long time to do that. I didn't even piss this morning when I got up so I would have a really full load. Yes sir, the pause that refreshes.'"[1]

The following is a documented account of the airdrop of Merrill Sauls, by another member who jumped:

"At 10:00 AM on March 24, 1945, a literal "sky train" of two and one-half hours in length flew in 226 C-47 and 72 C-46 transport planes carrying 17,122 paratroopers together with 906 gliders being pulled by 610 C-47 tow planes. Paratrooper Robert L. Wilson jumped from the lead serial of planes in this largest airborne assault in military history. He landed in the designated drop zone near Wesel. There, with others in "Branigan's Bastard Battalion,"—the 464th Parachute Field Artillery Battalion—he made history as part of the first artillery airborne outfit to land, reassemble, and fire their howitzer's east of the Rhine River. Field Marshall Montgomery, who had elaborately orchestrated this airborne assault, gathered Prime Minister Churchill together with Generals Eisenhower, Ridgway, and Brereton on the western bank of the Rhine to watch the air spectacle unfold. The objectives of this airborne assault—Operation Varsity—were achieved predominantly through the efforts of the U.S. 17th and British 6th Airborne Divisions. These troopers closed in along the Rhine all along its length to prevent any German stronghold from being secured on its bank. They later thrust across Germany's northern plains en route to Berlin. In sum, Varsity's success accelerated the Nazi collapse."[2]

When Merrill Sauls jumped near the town of Wesel, Germany (at the confluence of the Lippe and Rhine Rivers), he had on his person three hand grenades, C-rations, a carbine, a small map, a compass, and nothing else.

They also dropped separately a .50 caliber machine gun that landed about 75 yards away from the troops, on a farm area with an outcropping of trees.

Merrill crawled over to get the .50 cal going so as to counterattack against Germans who were raining intense fire down on them from the farmhouse. Once the Americans got the .50 cal working, they hosed down the farmhouse to quell the German attacks.

Sauls remembered that many Americans who had the misfortune to be caught up with their parachutes hung in trees were killed almost immediately by the furious German fire. German machine guns often cut the men in half as they hung helplessly.

A history of the 17th Division shows that at the end of the *first day* of the jump, some 2,873 Germans had been captured and hundreds killed. The cost to the 17th Division was considerable: 393 killed, 934 wounded, 164 missing.[3]

Once Patton got the tanks across the Rhine, the American troops were, in Sauls' words, "Dang glad to see them." After they linked up with the tanks, Merrill on day 3 after that was shot in the lower back (March 25th). He stated he lay in a horse trough for a long time before getting help.

He said he also once observed overhead a dogfight between an American P-51 and a German Luftwaffe fighter, with the American pilot getting the best of the German pilot.

It was very difficult for the American doctors and nurses to administer medical assistance to the wounded, as they were constantly under heavy bombardments. Sauls recalled "...a tent some 200 yards long, with cots three abreast, where one doctor and two nurses tried to treat the wounded. Every time they started treatment, the Germans attacked again, so they ran for the bunkers." He later was flown to France where he was treated for his wound for a month, a wound that never would heal properly. They finally packed his wound one last time, then sent him back into duty.

After the surrender of Germany, Sauls was assigned to patrol duty with another soldier on a jeep in an area near a power plant to ensure the plant manager kept the power on for the citizenry.

On one patrol, a German woman ran out to stop them, waving her arms wildly, and yelling to them that her son was a Prisoner of War in

America, and wanting to know how he was being treated. Merrill assured her as best he could that her son was being well treated by the Americans.

The German woman seemed satisfied, thanked him profusely, and brought Sauls and his buddy a cake. She also promised if they would stop every second week on their patrols, she would give them a cake each time, and she did.

The 464th Parachute Field Artillery Battalion troops eventually ended up in Munich, and, after a week or two, they were ordered to pack up and be ready to head out to fight in the anticipated invasion of Japan. After about two days at sea, the announcement was made over the ship's intercom that "Japan has surrendered, the war is over."

At the time of that announcement, the soldiers were eating in the ship's mess, and, upon hearing the good news, flung their food trays all the way to the ceiling!

Sauls was eventually sent back to Fort McPherson, Georgia, but he didn't have enough points to get out of the Army. Yet, after a 30-day leave, and a lessening of the Army's points system after the war's end, he then had enough points to get out. So, he headed back to Rose Hill and worked as a mechanic with his brother, W.A. Sauls, Jr.

Merrill later worked for the U.S. Postal Service for 24 years and married Correne Cain, of Beulah community, near Opp. Correne taught school for some 38 years, including stints at rural county schools Sanford and Blue Springs, as well as in the Opp City Schools.

They later moved north and Correne also taught school in Wilmington, Delaware, for some years, before they eventually moved back to the Opp area.

There they first built a house on the Ellis Road area, then later built a house on W. Gunter Ave where he now resides.[4]

JOHN CLUCK, JR.

Army Photo (Cluck's Personal photo collection)

John Cluck, Jr. was born April 3, 1923, in Georgetown, Texas, a small town just north of Austin. His father, John Cluck, Sr. was a cattle trader specializing in Jersey cattle. His mother Ida (Taylor) Cluck later in life had a credit bureau.

He attended the University of Texas, Austin, and worked toward a degree in Mechanical Engineering. Lyndon B. Johnson had helped him get an appointment to the Naval Academy at Annapolis, but his family didn't have the $100 needed to get him there for a physical examination (they had told him he had an "incomplete hernia" which required another examination before he could start at Annapolis).

So, rather than a career in the Navy, he attended classes at Texas, where he had no car and just rode a bicycle around campus. He worked in the school library to earn money so he could continue with his classes and earn that Engineering degree.

When World War II broke out, Cluck wanted to join the armed forces, but engineers weren't being accepted. So, he got a friend to get him a blank draft card which he filled out and then managed to get into the Army without letting them know he was an engineer.

He first attended some technical training as a Private at Yale University, then got into Cadet training. Cluck attended Cadet Technical Training at Lowery near Denver, Colorado, and finished with a commission as a second lieutenant. Given options, he decided to become a flight engineer and headed out to Salina, Kansas for aircrew training.

There, he qualified as a flight engineer on B-29 *Superfortress* bombers, the first and only airplane he ever flew aboard. Cluck ultimately completed 35 combat missions over Japan, hitting targets over Tokyo, Osaka, Kure and other locations.

Cluck was assigned to the 877th Bomb Squadron, 499th Group, 73d Bomb Wing, which was part of Twentieth Air Force.

He and his other crewmembers thought they probably would go home after 25 combat missions, as had been the case with the aircrews in the European Theater. However, when he completed his 23rd mission, the powers-that-be decided that crewmembers had to complete 30 missions before going home.

Then, after he had completed 29 missions, the higher-ups told the aircrews that, due to heavy losses, they had to extend it to 35 missions before they could go home.

Japan's industrial cities, with their concentration of wood and plaster buildings, were exceptionally vulnerable to incendiary attack. In a change of tactics (from high altitude to low altitude attacks), General Curtis LeMay sent his B-29s over Japan at night with devastating results. On March 9, 1945, some 15.8 square miles of Tokyo were burned out. Raids on Japan's six most important cities followed, and 105 square miles in the cities were destroyed, leading to the collapse of Japan's ability to make war, even before the atomic bomb was dropped.[1]

The B-29 missions were hazardous. On May 26, 1945, some 464 B-29s were dispatched from the Marianas to bomb Japan. Over five percent were lost over the target.

Twentieth Air Force losses were large in the war. Overseas, 512 B-29s were lost, with the deaths of 576 fliers and 2,400 missing. Many aircrew members who bailed out over Japanese territory were killed by irate Japanese civilians.[2]

Cluck completed his 35 missions without being shot down, and he and his crew were awarded the Distinguished Flying Cross. He served his obligatory time in the Army Air Corps, but left the service after the war ended.

Post-military, he became a Plant Engineer in Texas with Lufkin Foundry and Machine Company. He worked on oil pumpers that draw oil from the ground (affectionately called "Iron Horses" by those in the Texas oil business). He later became Chief Engineer of the company's Trailer Division.

Cluck has been interviewed on local television stations such as WSFA during specials about our military veterans. He also has studied our nation's currency and has written a "white paper" about how our currency is continually being devalued. He occasionally gives talks to different groups who are interested in that subject.

At the time of this interview, Cluck was at a meeting of VFW Post 96 in Montgomery, Alabama, accompanied by his son, Myron, who also served in the National Guard during the Vietnam War era. Myron Cluck was a computer repairman.

John Cluck and wife Martha Jo had two others sons, Don and Ed. All the Clucks live in Montgomery, Alabama.[3]

John Cluck passed away on December 1, 2013, after a lifetime of service to our country.

HUNTER "BILL" MAXWELL BISHOP, JR.

"He came home, and only 8 of the original 300 men in G-Company, 505 Parachute Regiment can say that."

Sergeant Hunter "Bill" M. Bishop, Jr. (Army photo)

Sergeant Hunter "Bill" Maxwell Bishop was born Oct 2, 1922, to parents Hunter M. Bishop Sr. and wife Lillian Evans (Logan) in Aucilla, Florida, an unincorporated community in Jefferson County.

After the war, he married Evelyn Lett. Their children were Raymond, Bruce, and Tonni.

He first entered military service at age 17 (1939), as part of Company M, 12th Infantry Regiment, 31st Division, Tallahassee Florida National Guard, and was there for about a year and a half.

His active duty began 25 Nov 1940. In WW II, Bishop was a member of Company G, 505th Parachute Infantry Regiment, 82nd Airborne Division, one of America's most storied units.

Bill Bishop was quite the soldier, a squad leader during WWII in Africa, Sicily, Italy, and jumping into Normandy ahead of the D-Day landings that came on June 6, 1944.

When it was all over his decorations/citations included Purple Heart, Distinguished Unit Badge, European-Africa-Middle East Medal with 3 Bronze Stars, Parachute Badge, WWII Victory Medal, and other awards.[1]

I first came to know about Bill Bishop through his nephew, Mike Bishop, and son, Bruce, who attended Troy State University with me in the late 1960s.

Mike pointed out an article in a newspaper about Bishop's service that had the following interview with WALB-TV of Albany, Georgia:

May 26, 2004 Cairo—An adrenaline rush—that's why World War II Veteran Bill Bishop says he wanted to become an Army paratrooper. The thrill of the fall landed Bishop in the middle of the most intense fighting during the war in Europe.

"I was in G-Company, 505th Parachute Regiment." Bill Bishop was one of the Army's elite paratroopers in World War II. "Our main purpose was to make the enemy give his life for his country. And, that's what we did."

His first combat jump came 9 July 1943 during the battle of (Gela) Sicily. "A 15-knot ground wind is considered dangerous. And, it was at a 30-knot ground wind." "Everybody hit hard. We had a lot of injuries. I got knocked out on the jump." But he came to and started fighting. "Bullets were flying everywhere. After we hit the ground, we could be in action in less than five minutes time."

The parachute soldiers would jump directly onto the enemy. "Jumping on top of them like that, they were just as confused as we were." The element of surprise made the regiment highly successful, but it suffered an unusually high number of casualties. "You're dog-gone right I was scared, and everybody else was scared."

Their objective in *Operation Husky* under Colonel James M. Gavin was to parachute behind enemy lines into an egg-shaped area near Gela, then close off roads leading to beaches and secure the drop zone for further use. Especially important near the DZ was enemy-held Objective Y—a series of 16 concrete pillboxes from which German gunners controlled movement on nearby roads.

The Americans took out the pillboxes and, despite being outmanned and outgunned, blocked elements of the Herman Goering Panzer Division from advancing toward the Allied landings on the beachhead.[2]

Bishop recalled two Navy ensigns jumped with one team to direct Naval gunfire. One ensign broke both legs, could not get out of the way, and was crushed by a Panzer tank.

Because of weather conditions, only 15% of the combat team had been delivered to the correct DZ. Yet, small groups of lost paratroopers cut every phone line that they found which devastated the communications of the Axis forces.

In a newspaper article about his service, Bishop talked about the stress of the paratroopers having to bury their own dead because there were no burial detail personnel to do the job. "...we buried our own and did the best we could with them. There were a lot of us that had to go to the Chaplin and have a little talk, that's the way it goes."[3]

Bishop remembers flying into Normandy on D-Day, sticking his head out the plane's door and seeing a sight to behold. "It was a fabulous sight to see. All the aircraft and ships going in one direction; I wasn't scared."

Bishop's mission in Normandy was to jump into a town near the shore (Sainte Marie Eglise) and cut off the Germans headed to the beach, but the jump went bad. "I knew something was wrong. Come to find out, I had blown nine panels out of my parachute."

Bullets pierced his chute and he slammed into the ground, but survived. "I drug myself off the field, with a femur sticking out. I pulled into this drainage ditch. I laid out my weapons, my grenades and things, and got ready—like Custer's Last Stand."

But, Germans soon found him. "I thought they were going to kill me, but they didn't." Amazingly, they only took his weapons and left. "They came back an hour later, picked me up and carried me to their aid."

Another account of other 505th paratroopers in action at Sainte-Mere-Eglise on D-Day was described as follows:

"The early landings, at about 0140 directly on the town, resulted in heavy casualties for the paratroopers. Some buildings in town were on fire that night, and they illuminated the sky, making easy targets of the descending men. Some were sucked into the fire. Many hanging from trees and utility poles were shot before they could cut loose. The German defenders were alerted.

"A famous incident involved paratrooper John Steele of the 505th PIR, whose parachute caught on the spire of the town church, and could only observe the fighting going on below. He hung there limply for two hours, pretending to be dead, before the Germans took him prisoner.

"Steele later escaped from the Germans and rejoined his division when US troops of the 3rd Battalion, 505 Parachute Infantry Regiment attacked the village, capturing thirty Germans and killing another eleven. The incident was portrayed in the movie *The Longest Day* by actor Red Buttons. Later that morning, about 0500, a force led by Lt. Colonel Edward C. Krause of the 505th PIR took the town with little resistance. Apparently the German garrison was confused and had retired for the rest of the night. However, heavy German counterattacks began later in the day and into the next. The lightly armed troops held the town until reinforced by tanks from nearby Utah Beach in the afternoon of 7 June."[4]

He (Bishop) was a prisoner of war, but not for long. "The 4th Infantry came in and overran their position." The Germans fled from the U.S. soldiers, leaving Bishop behind. "I couldn't have thanked them enough, even though they were the enemy." He spent more than a year in the hospital. "I wound up with an 11-inch plate in my femur with six screws."

But he came home, and *only 8 of the original 300 men* in G-Company, 505 Parachute Regiment can say that. Bishop was honored with two Purple Hearts for his bravery and service in World War II. He now lives in Cairo (Georgia) with his wife (end of WALB-TV 2004 interview).

After the war ended, Bill Bishop and his brother, Ralph (Mike Bishop's dad), went to Horology School—watch and jewelry repair—in Tampa on the GI Bill, and both opened successful jewelry stores which they operated for their entire careers. Bill sold his Cairo jewelry store to his nephew, and the store is still in operation as of today, with Mike Bishop's wife, Diane, as store manager.

More than seven years after the above account of his World War II experiences, Bill Bishop passed away on February 18, 2011, and is buried in Cairo, Georgia. He may be gone, but he left a legacy of service to his nation, a legacy of which those in his family are extremely proud.

Like so many families of that era, others in Bill's family served as well. His brother (and business partner) James Ralph Bishop served in the Navy from December 2, 1943 until April 14, 1946. He was in the Mediterranean and Normandy Invasion campaigns.[5]

Having served their country in uniform, Bill and Ralph worked together to serve the citizens of Grady County though their excellence in the jewelry/watch repair business.

I like to think of them as *just ordinary citizens who performed extraordinary service for their nation, in its time of need!* The Bishops exemplified the core value of "Service Before Self," a core value no doubt that is shared by our Army, Marine Corps, Navy, Air Force, Coast Guard, and so many public service organizations of our country.

Bill Bishop at Fort Bragg, 1943 (family photo)

JOHN PATRICK "JACK" O'BRIEN

"He ripped a page out of a school book depicting China's East Coast"

Cadet Jack O'Brien 1943 (his family photo)

John Patrick "Jack" O'Brien was born November 4, 1924, in Jersey City, New Jersey to parents Clarence and Alice (Hansen) O'Brien.

As a youngster, Jack's schools, both grammar and high school, were Catholic (St. Michael's High in Union City, New Jersey).

In his youth, "sports was everything" in the young lives of Jack and his contemporaries. He participated in football and baseball, and had a particular fondness for hockey. In fact, he was at a hockey game in New York City on December 7, 1941, and remembers vividly the scene.

That afternoon, at the end of a period of play, all the lights went out, a floodlight came on and illuminated the American Flag, and a sign was flashed up saying "The Japanese have raided Pearl Harbor. All military personnel report back to your bases." He recalls, "You could have heard a pin drop for the next two or three minutes," as those in the arena gathered their thoughts and composed themselves.

At age 18 he joined the Army Air Corps. On August 3, 1943, his parents dropped him off at the Newark railroad station and bid him good luck for his basic training at Greensboro, N.C. Afterward, he was off to start Aviation Cadet training at Maxwell Field in Montgomery, Alabama.

There, he flew primary aircraft such as the *Stearman* in the early stages of cadet school, and eventually finished up his training at navigator school at Hondo, Texas. Afterward, he served as an instructor there for several months.

As a second lieutenant, he deployed to the Pacific Theater of World War II in the latter phases, arriving in China in May, 1945, where he was flying aboard C-47s at Chunking as part of the 27th Troop Carrier Squadron. Their mission was airdropping troops or supplying them with food, ammunition, and other supplies.

U.S. LTG "Vinegar Joe" Stillwell had retreated from Burma in 1942, trained Allied troops, and re-entered to drive out the Japanese. By the spring of 1945, British General Orde Wingate's *Chindits*, Brigadier General Frank *Merrill's Marauders*, Claire Chennault's *Flying Tigers,* and Stillwell's improving troops had all played havoc with the Japanese in some of the most vicious jungle warfare in the 20th Century. O'Brien recalled he dropped supplies to jungle fighters in Burma, but they were not always sure which unit they were.

Note: The 27th Troop Carrier Squadron is now an inactive United States Air Force unit, but its last assignment was with the 443d Troop Carrier Group, based at Liangshan, China. After WWII ended, it was inactivated 27 December 1945.[1]

With the 27th TCS, O'Brien never flew "the hump" as did many other C-47 crew members in the C-B-I Theater, but his crew made a great many paratroop drops departing from their mountainous airfield, then dropping near or behind Japanese targets in relatively flat and lowland terrain. When they dropped Chinese troops, they always had an American Jumpmaster.

He remembered one particular mission where they dropped all Chinese paratroopers, and the Japanese evidently had figured out where the drop would be. His crew got word later that all the Chinese troops had been killed, either cut to pieces by Japanese machine gun fire in the air, or overrun and killed once on the ground.

His tensest moments were usually returning to their mountainous area airfield when the weather would deteriorate, making it testy to get the aircraft on the ground. Without oxygen aboard the aircraft, all flights had to be at or below 10,000 feet, so there were added elements of danger.

O'Brien's crew had one particularly significant mission when on August 27, 1945, they flew to Mukden, Manchuria, to pick up and return American and Allied Prisoners of War after their liberation. Navigating was very rudimentary with the basic instruments aboard the C-47s. It was so basic that on this occasion, he just used a school map—he ripped a page out of a school book depicting China's East Coast and a rail line that originated at a seaport. He and crew simply followed the coastline until they spotted the railhead, then turned north and followed the rail line until they found their destination.

When they arrived at the airfield, lined up on both sides of the runway were about 15 new C-47's, no paint, with the Russian stars on the wings. There were also many *Aircobra* fighters. Jack's 27th squadron C-47's had loads of miles on them, but the Russians had brand new craft. He and crew were escorted down by three Russian fighter planes. They found out later two of the escort pilots were women.

Among those recovered were American Army General Jonathan Mayhew "Skinny" Wainwright, who had surrendered American forces at Corregidor, Philippines, on June 9, 1942, as well as British Lieutenant General Arthur Ernest Percival, who had been Commander of Singapore until its surrender to Japanese forces.

At the time, Singapore's surrender to the Japanese was the largest capitulation in the history of the British Empire. O'Brien and crew also recovered a Dutch general, a governor of Sumatra, and other generals, majors, sergeants, and even a private first class—a total of two dozen who had been detained as prisoners by the Japanese.

Wainwright had been held for three years in prison camps in northern Luzon, Formosa, and Liaoyuan (then called Xi'an and was a country *within* Manchukuo), but was liberated by the (Russian) Red Army in August, 1945. He later was awarded the Medal of Honor, over the protests of General Douglas MacArthur.

O'Brien said it was obvious from their physical appearance in Manchuria that the prisoners had been severely mistreated while in captivity. Pictures verify that fact, too.

He and his crew later discovered they had flown some missions under OSS orders (the OSS was predecessor of today's CIA). An OSS agent coordinated the POWs' release.

After the liberation of the American POW's, a subsequent meeting of General MacArthur and LTG Wainwright occurred at the New Grand

Hotel, Yokohama, Japan, on August 31, 1945. Wainwright was aboard the USS *Missouri* to observe a formal surrender of Japan on September 2, 1945, and so was British General Percival. Official photos show they stood behind MacArthur.

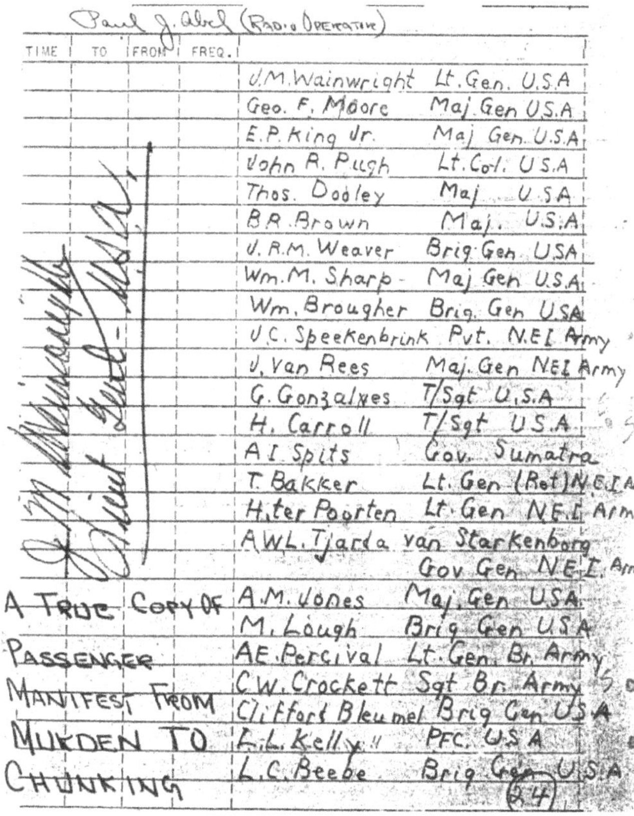

Paul J. Abel (Radio Operator)

TIME	TO	FROM	FREQ.		
				J.M. Wainwright	Lt. Gen. U.S.A
				Geo. F. Moore	Maj. Gen U.S.A
				E.P. King Jr.	Maj Gen. U.S.A
				John R. Pugh	Lt. Col. U.S.A
				Thos. Dooley	Maj U.S.A
				B.R. Brown	Maj. U.S.A
				J.R.M. Weaver	Brig. Gen. U.S.A
				Wm. M. Sharp	Maj Gen U.S.A
				Wm. Brougher	Brig. Gen U.S.A
				J.C. Speekenbrink	Pvt. N.E.I Army
				J. Van Rees	Maj. Gen NEI Army
				G. Gonzalves	T/Sgt U.S.A
				H. Carroll	T/Sgt U.S.A
				A.I. Spits	Gov. Sumatra
				T. Bakker	Lt. Gen (Ret) N.E.I Army
				H. ter Poorten	Lt. Gen N.E.I Army
				AWL. Tjarda van Starkenborg	Gov Gen N.E.I. Army
				A.M. Jones	Maj. Gen U.S.A
				M. Lough	Brig Gen U.S.A
				A.E. Percival	Lt. Gen. Br. Army
				C.W. Crockett	Sgt Br. Army
				Clifford Bleumel	Brig Gen U.S.A
				L.L. Kelly	PFC. U.S.A
				L.C. Beebe	Brig Gen U.S.A

J.M. Wainwright, Lieut. Gen. U.S.A.

A True Copy of Passenger Manifest from Mukden to Chunking

(24)

Passenger Manifest with Gen. Wainwright first passenger listed
(Courtesy of Paul J. Abel, Radio Operator on Jack O'Brien's aircraft)

As all crewmembers prepared to come back stateside, O'Brien volunteered to fly as a navigator rather than take a ship as most did, saving himself eighteen days getting home. Jack O'Brien left active Army duty as a first lieutenant in early 1946, then served in the inactive reserves enough years to get another sixteen-and-a-half "good years" to qualify for retirement benefits in the grade of major.

Along the way, Jack married his sweetheart, Virginia Murphy, from Roanoke, Virginia. They had two sons, Pat and Dan. I met Dan's wife and Jack's granddaughter, Meghan, in Rome, Georgia, at Berry College horse shows.

Jack's beloved Virginia passed away in 2008, after they were married for 55 years. As Virginia's health deteriorated in her last years, Jack took up painting as a way to relieve the stress of caring for a loved one with dementia. He continues today with this hobby that includes painting scenes of landscapes, shrimp boats, musicians, and the human body.

Jack O'Brien lives in Mount Pleasant, South Carolina, enjoying life with his paintings, watching his granddaughter, Meghan, ride horses, and getting with his many friends to reminisce about the adventures and close calls he had in World War II. [2]

He is a very patriotic individual who considered it a privilege to serve his country in its time of greatest need. A grateful nation appreciates Jack O'Brien and his aircrew as well.

EVON WILLIAMSON

"A German opened up on him with a 9-mm 'burp' gun"

Evon Williamson was born September 24, 1922 in Covington County, Alabama, to Irvin and Mary Beulah (Bozeman) Williamson who were share-cropping farmers. Williamson picked cotton, stacked peanuts, hauled hay, pulled corn, and assisted with the family's gardening in order to help the family survive. His brothers—James, Clayton, and Madison—were all in the military, and he had a sister, Nell.

Growing up not far from Gantt, the family moved around, and he completed some grades at Thurston School in Crenshaw County, about 15 miles from Dozier. He later attended Andalusia High School, and was drafted by the Army in 1943 while he was finishing his junior year.

Williamson was sent by train for basic training at an airbase near Portland, Oregon. When he left there, the new Private was sent to New York from which he departed for overseas in a Quartermaster job, leaving by ship from Staten Island, N.Y. for Africa.

After arriving North Africa near a place called Oran (at a time when the British-American assault on Vichy French North Africa was ongoing under *Operation Torch*), the Army decided he was needed in Infantry more than in Quartermaster. In Africa, he completed 18 weeks of Infantry training as a machine gunner before heading into combat while assigned to the 351st Regiment of the 88th Division, which came to be known as "The Fighting Blue Devils."

After his first night with his Infantry unit, his first combat was in Italy the following day and lasted until his 21st day of hostilities during one of three of the 88th's campaigns. Williamson was part of the 88th's Po Valley Campaign.

He remembered the gruesome effects of all-out combat. He recalled once when an American soldier left his foxhole and made a mad dash, an artillery round struck near the soldier, blowing him apart, with one of his legs falling back into the foxhole he had just departed.

On the 21st day, he and other American troops had seized a 4-story building not far from a German command post. The Germans surrounded the building and demanded the Americans surrender.

One of Williamson's buddies spoke German and went outside and surrendered. Williamson stayed put until they tossed in a concussion grenade, broke down the door, and began shooting. A German opened up on him with a 9-mm "burp" gun, and he was struck several times in the left hand and right knee before being taken prisoner on October 21, 1944. One of those bullets remained in his knee for 37 years, and, to this day, he has suffered *osteomilitis* from his wounds.

As the Americans advanced, the Germans often moved Evon and the other prisoners. He ended up in a POW camp in Austria. His mother received an initial War Department letter dated October 25, 1944, signed by Major General J.A. Ulio, Adjutant General, with the news that Evon was "missing in action and presumed dead." Three months later, his mom got the first letter from him reassuring her that he was alive and well and a prisoner of war.

As a POW, he said he was not questioned by the Germans, nor was he ever tortured or mistreated, other than not having enough food and having no medicine. He could recall only once getting a Red Cross package. He also said for many months he was the only American in his camp. The others were British. The second American prisoner to arrive was a black American soldier who said he was a pilot.

His camp was very near a much larger prison camp holding Russians captured by the Germans. He recalled there were Russian women and children as prisoners in those camps as well, as it was not unusual for the Russians to carry their dependents with them into combat. The Russian wives cooked for their husbands in the camps. A few spoke English, and he remembered them as being "very backward, with little to no education."

Williamson said once while he was in a hospital, German doctors brought Fascist Italian dictator Benito Mussolini in and asked if any of the POWs could identify the man as Mussolini. He declined, since he wasn't sure what Mussolini looked like.

Near the end of his captivity, his guards deserted the camp. He by this time had shed a leg cast and was able to hobble to a nearby town with two other captives. Nearing the town, he told the other two, "I hear the sound of a Jeep." They found the Jeep, flagged it down, and it was being driven by reporters of the *Stars and Stripes* newspaper. The reporters identified the Americans as POWs and arranged for an aircraft to fly them out the following day to Algiers.

During a visit from a USO Camp Shows, Inc. group that entertained the troops, a girl named Irma Attridge asked Williamson and two of his buddies to draw cards to see who would be the beneficiary of her drawing a photo of one of the three. Evon drew the high card, and the below drawing (dated June 1, 1945) was completed in some 15 minutes on a brown paper bag by Ms. Attridge.

Williamson spent one night in Algiers, then was flown to Italy to the 21st General Hospital where he spent 4-5 days being treated. He subsequently was flown to Miami, Florida, grateful that he was getting back to the good old USA! He arrived in Miami just a few days before the Japanese surrendered in the Pacific.

After Miami, he was sent to an Army hospital in Augusta, Georgia. From there he was transferred to Northington General Hospital in Tuscaloosa, Alabama, where he spent several months convalescing. While there, he managed to get passes to travel to Andalusia to see his folks as well as a special lady, Maxine Hildreth, from the Pleasant Home community.

In January, 1946, Williamson was discharged from the Army with 100 percent medical disability. He would not be able to bend his leg where he took the 9 mm bullets to the knee, and his left hand is disfigured and has very limited function.

Later that year, he and Maxine were married, and they have three children, Gayle, Jennifer, and Jeffery. They also have four grandchildren and four great-grands.

In 1981, a Doctor Bowness came to Andalusia and examined Williamson's hand and leg which seemed to develop frequent infections around the knee. He convinced them the bullet should be removed, as opposed to what VA doctors had recommended, so they removed the round, and there have been no more infections from the bullet that had been lodged for 37 years.

Evon's older brother, James, was also in the Army in the WWII era, serving with the 36th Infantry Division in Europe. Another older brother, Clayton, was in the Army serving in the Pacific. A younger brother, Madison, served in the Army in the Pacific, but after the war was over. His older sister, Nell, married Claude Tankersly.

Maxine's brother, James Curtis Hildreth, served 33 years in the U.S. Air Force in Food Services. His last three years were at Eglin AFB, Florida with the Air Force Inspector General team that traveled to different bases on major inspections.

It was difficult to raise a family on military disability pay in 1946, so Maxine worked many years while raising a family. She worked 20 years in a textile mill, worked in a pecan plant, put herself through Nursing School and worked 20 years in a doctor's office, and then worked 20 years in Home Health Care. [1]

A remarkable and strong woman she is, as are so many spouses of disabled veterans, and she deserves so much credit for insuring Evon's well being while working and taking care of her children.

When he was mobile enough, Evon's hobbies included hunting dove and deer, as well as fishing. He organized a big hunting club which many others enjoyed.

America owes a debt of gratitude to the Williamson family, as Evon and Maxine and so many others in the extended family contributed so much to our country's well being. A grateful nation thanks them for their many sacrifices.

Drawing of PFC Williamson by Irma Attridge, USO CAMPTOURS

LEE JACK LOWERY

"He eventually was to become an Armorer-Turret Gunner"

Lee Jack Lowery in Marine Uniform (Courtesy of son Keith Lowery)

Marine Sergeant Lee Jack Lowery was born July 3, 1923, in the tiny community of Brushy Creek near Georgiana, in Butler County, Alabama, only a stone's throw from the birthplace of Hank Williams, Sr. In his youth, he worked on his parents' farm doing all the usual chores young boys did in the 1930s and 40s. The family got by on a meager income.

He graduated from Georgiana High School in 1942, and, on February 22, 1943, he and his younger brother Murray Lowery traveled to Birmingham, where both enlisted in the service – Jack Lowery in the Marine Corps and Murray in the Navy.

Lowery completed his basic training at the Marine Corps recruit training station in San Diego, California, where he experienced all the hellish boot camp events that the most sadistic Marine Drill Instructors could dish out.

Following basic training, he completed weapons training at NAS Norman, Oklahoma, June-July, 1943, and from there he was off to Naval Air Gunners School in Purcell, Oklahoma, August-September. By October, Lowery was at Cecil Field in Jacksonville, Florida, where he completed ground school for radio, radar and code, blinkers, flags, aircraft recognition, and other subjects.

November and December of '43, he started flying with an aircrew where they trained for fixed- and flexible-gunnery runs and dive bombing, then he got a short break where he went home on leave back at Brushy Creek for a few days at the end of December.

With all his training, he eventually was to become an Armorer-Turret Gunner on a Marine version of the B-25, classified by the Marines as a PBJ-1H bomber.

In January, 1944, he trained at MCAS Cherry Point, N.C., assigned to the 613 Bombing Squadron-Training. Following, in February, he was trained at the Naval Air Station at Boca Chica, Florida, with more training runs.

In March, his CO used 10 aircraft, 10-men-per-plane, and flew everyone to Havana, Cuba, on liberty, which was a treat for Jack and the others.

In November, 1944, Jack boarded a ship at San Francisco, with destination somewhere in the Marshall Islands, via Pearl Harbor.

At Pearl in December, he met his brother Murray for just a few hours on a short stop when they were able to shuttle between ships using a small boat. No shore leave was allowed for either Marines or Navy personnel, so after the quick visit, he continued sailing for final destination.

Arriving late December, he was stationed in the Marshall Islands (Kwajalein) until late 1945 and assigned to Marine Bombing Squadron 613 (VMB-613). Kwajalein had earlier been the site of some of the heaviest fighting in the Pacific. By the time Jack's ship arrived, the troops had been on the high seas for 32 days. By then he had overcome any early sea sickness, acquired his "sea legs," and tolerated the trip in fine fashion. At long last, after so much training, and so many different locations, he was ready for combat.

Lowery's squadron aircraft escorted convoys of Allied and Navy vessels and attacked Japanese targets at every opportunity. His job as a gunner was to protect his craft against airborne attacks by enemy fighter aircraft such as the Japanese *Zero* or dive bombers.

Escorting convoys of Allied ships was the everyday business of VMB-613. His squadron patch included a rather large skull, obviously indicating their purpose in life was to kill the enemy.After the Japanese surrendered, Jack was returned to the States via ship in time to be home for Thanksgiving, 1945. After the war, Lowery settled back in the Brushy Creek Community, married his sweetheart, Dorothy (Wingard) Lowery, on Christmas Day, 1946, and proceeded to farm and raise a family.

His contributions to our nation continued through raising cotton to clothe our people, and foods such as corn and peanuts to feed them. Farming was his life in his early years, and it would be his life after the war. He experienced true happiness in what he was doing.

Jack Lowery was not one to brag about his exploits during World War II. In fact, he rarely mentioned anything at all about his role in the war to his son, Keith, or to either of his daughters, Melinda and Amelia. Keith just stumbled upon a photo one day while surfing the internet and was astonished to find his father's picture with other members of his crew in VMB-613.

Jack passed away in March, 2003, and Dorothy passed away on February 1, 2013.

Continuing in the family traditions, Jack's son, Keith, enlisted in the Marines in 1969, went through basic training at Parris Island, and additional training at Camp LaJuene, N.C. Nothing had changed since his dad went through basic. The DI's were just as sadistic as ever, turning boys into men who were forged in the fire of stress and physical fitness.

Keith was in the Marine Corps Reserves from 1969-1975, was categorized as an "0-111" (basic "grunt") in Infantry for a while, then was with the Supply Company in a Montgomery, Alabama, reserve outfit. [1]

Keith is a graduate of Auburn University, with a degree in Engineering. His life's career was in the Environmental Field, working until his first retirement with the Alabama Department of Environmental Management (ADEM) in the Water Department.

He also had his own consulting business, contracting with Lanier Environmental Consultants to work at Maxwell AFB, AL, as well as Keesler AFB, MS for many years.

Keith and wife Peggy (Reeves) from Tallassee, live in Montgomery, and he continues to serve the public well through private contracting with different towns and municipalities in the State of Alabama to keep their water supplies clean and safe.

NOAH NELSON COWEN

"The P-51 carried the name Margaret Two."

2Lt Noah Cowen with P-51 Mustang (photo from his family)

In discussing Alabama servicemen and women who fought bravely in WWII, Charles Saunders acquainted me with a photo he had made in Belgium of the grave of Noah Nelson Cowen, Army Air Forces pilot who gave his life in Europe and was never brought home.

Noah Nelson Cowen was born Feb 20, 1922, to Frederick Daniel and Susie (Ham) Cowen, in the farming community of Curtis, along U.S. Highway 84 between Opp and Elba, Alabama. According to Shelby Strickland Cowen, he lived along a row of houses with so many Cowens that it came to be referred to as "Cowen's Row." His house was number 286, and was in an area dominated by agriculture. Noah, as a youngster, did more than his fair share of chores, as did his brother, Frank Harris Cowen. The 1940 Census noted that he worked 44 hours per week and earned about $130.

Cowen entered military service as a Private on March 1, 1943, at Miami Beach, Florida, with two years of college.

Later, at Aviation Cadet training, he was an excellent student, and, after graduating as a young second lieutenant, was chosen to fly the Army Air Force's top-of-the-line P-51D *Mustang* fighter aircraft in World War

II. He lost his life only a few months after the D-Day invasion of June 6, 1944.

Assigned to the 350th Fighter Squadron, 353rd Fighter Group, operating from Raydon, England, Lt Cowen was shot down in Germany near Hohenhorst, around the Steinhuder Lake area (19 miles from Hanover) while flying P-51D (#44-14538) against several German FW-190s. Most all fighter aircraft had a girlfriend's or mother's name. His P-51 carried the name *Margaret Two*. Margaret must have been somebody special.

He and 1st Lt Alvin Lee Deeds, flying P-51D (#44-14727), were credited with one victory each before being shot down on 5 Dec 1944, over Hohenhorst, Celle, Niedersachsen, Germany. [1] Deeds' P-51 carried the name *Bee Jay*. Their A-Group call-sign was "Seldom," B-Group, "Persian."

It's possible Cowen had scant experience. War Department records from that era showed that some pilots entered combat in 1942 with less than one hour in their assigned aircraft. Another outfit, the 357th Fighter Group (Yoxford Boys) went to England in late 1943 having trained on P-39s, then flew P-51 *Mustangs*, never having seen a P-51 before.

Cowen's remains never returned. He rests in an American Cemetery in Neupre, Belgium, in the Ardennes area. [2]

Noah Nelson Cowen's marker in American cemetery, Neupre, Belgium (Plot B, Row 34, Grave #43)(Charles Saunders' photo)

Noah Cowen's decorations included Purple Heart, American Campaign Medal, European/African/Middle Eastern Campaign, and WWII Victory Medal.

THOMAS DWIGHT TEEL

"Flying in an open cockpit is like being near heaven."

Lt Dwight Teel (photo courtesy of brother, Tommy Teel)

A compatriot of Noah Cowen's—another pilot from Opp—Lt. Thomas Dwight Teel was similarly shot down and killed in WWII. Teel's mother, Tula, was a sister to Opp's Bill Ray, longtime employee of the Donaldson family that ran an agriculture center and cotton gin.

Teel had dated his sweetheart, Mildred Benton, but they put things off, and she later married J. L. Channell. The high school stadium is named for Channell's father. Teel had one younger brother, Thomas H. Teel, Jr., who was 17 years younger than Dwight and survives to this day, living in Valparaiso, Florida, having worked 41 years at Eglin AFB.

Dwight was born July 31, 1924, and graduated from Opp High School in 1942. His father was Thomas Howard Teel, Sr. Quite the athlete, Dwight had been quarterback on his high school team.

After high school, he had enrolled at Alabama Polytechnic Institute, now known as Auburn University. Aware that the draft was just around the corner, Dwight volunteered for the Army Air Corps, enlisting on March 10, 1943. He did his flight training at Helena and Newport, Arkansas, then further training at Craig Field near Selma, Alabama, where he received his pilot wings and a commission as a second lieutenant on December 5, 1943. He once sent a letter home saying, "Flying in an open cockpit is like being near heaven." He received further training and arrived in England prior to D-Day and would fly his first mission on D-Day. Amazingly, he flew six sorties that day!

Qualifying in the P-47 *Thunderbolt,* Teel was assigned to the VIII Fighter Command Station #369 at Martlesham Heath, England. He flew as number 4 of the "Vortex Red" section. Lt Col Philip E. Tukey, Jr., of Elizabeth, Maine, commanded his 356th Fighter Group.

The 356th Group was under the 67th Fighter Wing, and records indicate Teel was flying with the 360th Fighter Squadron. (Note: Martlesham Heath is 1½ miles from Woodbridge, a location with which this author is very familiar, having flown to RAF Base Woodbridge multiple times for training with the 67th Aerospace Rescue and Recovery Squadron which was there in the 1980s).

A hometown newspaper's description of Teel's summer of 1944 activities read, "Lt. Teel has been attacking German ground positions, bridges, railroads, and marshalling yards and chasing *Jerry* troops, tanks and trucks from the roads of France with withering machine gun fire from the eight .50-caliber guns the fighter carries." [1]

Things initially went well for Teel. Flying on a mission some 20 miles south of Hamburg, Germany, during a bomber escort mission, he was quoted as saying, "We were climbing to intercept about 28 Messerschmitts when I spotted an ME-109 on the tail of a -47. I immediately did a wing over and started after the *Hun*. Suddenly, my plane went out of control, and from 30,000 feet I spun down to 14,000.

"When I recovered, I noticed the *Jerry* was taking pop shots at me, and it was a most uncomfortable feeling. I looked up and there he was slightly to the right. I waited until he skidded in front of my gunsight and gave him a burst of .50 cal. machine gun bullets. I saw hits along the fuselage and wing roots. Smoke started pouring from the underside and the pilot jettisoned his canopy and bailed out. I did not see his chute open." [2]

The same article of September 14 that quoted Teel's account of his previously having shot down an ME-109 was headlined: Lieut. Teel Missing In Action. It stated the War Department had on Monday, September 11, notified Teel's parents on 407 West Avenue that their son was missing in action over France and Germany since August 29, 1944.

Tula and T. Howard Teel Sr. years later tried to determine as much as possible as to their son's fate. They were eventually able to correspond with Teel's wingman, a pilot named Williams from Tennessee, who wrote the grieving parents letters. Teel's aircraft, serial number 42-22524, carried the stencil *Grace.*[3] The wingman said it was hit by anti-aircraft fire at

10:45 in the morning, followed by Teel's subsequent bailout near Lille, France. The wingman observed that Teel was drifting down in what looked to be a good parachute.[4]

At the time, some pilots were being shot hanging in open chutes before they landed on the ground. Shooting a defenseless airman hanging in an open parachute is against the *Geneva Conventions* today, but at that time, "anything goes in war" was accepted by many.

A letter in French from some of those who found Teel's body arrived at the Teel home after the war. With no translators available, they asked the Air Force for help in translation. The letter was lost prior to the translation. With that bureaucratic "snafu," Howard and Tula's last, best chance of learning what actually happened to Dwight disappeared. They went to their graves without knowing.

Teel's brother, Thomas H., stated that a group of French buried Dwight's body in a small village church yard cemetery at Vitry-en-Artois, about 25 miles SSW of Lille.

In 1947, the U.S. Army moved his gravesite to the much larger Normandy American Cemetery at Colleville-sur-Mer, France, Plot A Row 9 Grave 38.

An American Battle Monuments Commission email of June 11, 2013, to Dwight's nephew, Chris Teel, indicated a Normandy American Cemetery Interpretive Guide, Mr. Constant Lebastard, had done much research and confirmed that village observers saw that Dwight Teel was indeed strafed and killed by a *Luftwaffe* pilot as he drifted down in his chute.

Teel's awards and decorations were: Distinguished Flying Cross, Air Medal with 3 Oak Leaf Clusters, and Purple Heart.

When he flew six sorties on D-Day, he was only 19 years of age. First Lieutenant Thomas Dwight Teel was one hell of a warrior who at age 20 gave his all to our country in its time of need! We salute his service and valor.

CAPT. WRITES OPP MOTHER OF SON'S DEATH

"I wanted an American to be buried under an American flag"

The following letter was received by Mrs. Lena K. Colvin, 16 Cole Street, from Capt. Joseph A. Archibald, III, somewhere in France.

Dear Mrs. Colvin,

I am the Captain of a U.S. Army tug, which has been stationed with the British in the British rector of the Normandy Beachhead.

On the 26th of June, the Captain of a British cruiser sent for me. He told me that he had recovered from the water, the body of your son, Pvt. Grady E. Colvin, and asked my advice on the matter. I told him to send the identification tag to the Adjutant-General's office in Washington. Then I told him that I would bury your son at sea because it was a question of either he doing it or my doing it, and I wanted an American to be buried under an American flag.

So at dusk on the 26th of June I pointed the bow of the ship toward America and with all the Military honors I could muster on a tug, I buried your son Grady in the English Channel.

I had to wait this long before writing because of War Department regulations.

Sincerely yours,
Joseph A. Archibald, III

Pvt. Colvin resided in Opp before volunteering for service. He was age 25 and received his education at the Westview School and was employed at the Opp & Micholas Mills until he entered the service. He has a brother now employed with the mills.[1]

Author's Note: Grady Colvin was buried at sea three weeks after D-Day. Grady's brother at the mills was Newt Colvin, whose son Sammy graduated high school with me in 1965.

KOREAN WAR

On June 25, 1950, North Korean troops with Soviet-made T-34 tanks and 122mm howitzers had swept across the 38th parallel, overwhelmed South Korean troops, forced South Korean strongman Syngman Rhee's government to re-locate south, and ultimately pushed down the peninsula toward Pusan, before the advance could be halted.

President Harry Truman proposed a "police action" to counter the aggression, the United Nations quickly voted to intervene with a multi-national force, and the vote was successful when the Soviet Union did not veto the action.

The U.S. ultimately would provide 88 percent of the 341,000 international forces committed to help South Korea repel the North Korean—and later Chinese—forces. Some 20 other countries joined the United States and South Korea.

General Douglas MacArthur first brought from Japan two U.S. divisions – the 24th and 25th – and they had little effect in halting the advance as "...the units began crumbling as fast as those of their new ROK allies... Defeatism crept into the high command. Correspondents wrote grimly of an imminent 'American Dunkirk.'"[1]

In a bold move, General MacArthur in September conducted the Inchon Landing 200 miles behind North Korean troops and cut them off from their supply lines, leading to the collapse of the North Korean Army. They beat a hasty retreat, and American and UN forces pursued them.

After fighting far north almost to the Yalu River and Chosin Reservoir where they encountered hordes of Chinese fighters, the Americans were pushed back south.

"When the People's Republic of China intervened in the Korean War in late November 1950, its 250,000 ground forces threatened to cut off and destroy UN units operating in the mountains of North Korea. To prevent that catastrophe and to concentrate UN units in more easily defended terrain further south, on December 9, General Douglas MacArthur, Commander in Chief, United Nations Command, ordered evacuation by sea of the U.S. X Corps."[2]

That evacuation took place from 9 through 24 December and resulted in troops from U.S. X Corps, 7th Infantry Division and 3d Infantry Division, plus the Republic of Korea I Corps, including the 3d and Capital Infantry Divisions being evacuated. The U.S. 1st Marine Division came out, too.

The units put out to sea on many vessels, then disembarked troops at safer locations further south and picked up the fighting (over 100,000 troops were moved, as well as more than 17,000 tanks, by Rear Admiral James H. Doyle's Task Force 90).

After the Americans and ROK re-grouped, fighting continued until the 1953 Armistice, after which an uneasy truce prevailed.

The fighting in Korea claimed 33,686 American battle deaths and 2,830 non-battle deaths, with over 92,000 injured. Some 17,730 died of "other" causes DoD wide. Some 8,176 went missing in action, including POWs. Losses for South Korea were more horrendous: 137,899 military deaths, and 373,599 civilian deaths.[3]

The Korean War was different than previous American wars, such as WWII and WWI, where anything less than total victory with unconditional surrender was unacceptable. "The experience was unique in three ways. First, it was our first limited war in modern (twentieth-century) times. As such, it represented a discontinuity in experience that required painful learning and adjustment. Second, it was the first significant Cold War confrontation and represented a novel challenge to the emerging American role in the world. Third, it was the first major American military engagement not preceded by a formal declaration of military intent by the Congress of the United States..."[4]

Fighting a limited war would present problems for many in the military, from General Douglas MacArthur all the way down through the ranks to those in Korea on the front lines for as many as 18 months (such as Kurt Miller who believed MacArthur's axiom that *there is no substitute for victory*, and whose disdain for Harry Truman is in his story that follows).

Total war generals and *limited objectives* politicians make for a bad mix. MacArthur's underestimating China's willingness to fight, as well as his desire to pursue total war when they were repelled, led to his dismissal by President Truman who wanted nothing to do with an Asian land war beyond a UN mandate.

That firing likely led to Truman's decision not to run for re-election after he was severely criticized by many Americans who so revered MacArthur that in many quarters he was considered an "American Caesar."

MILLARD E. MARTIN

Pvt Millard Martin (photo courtesy of Martin family)

Millard E. Martin was born on 26 October 1931, to Millard Fillmore and Flossie (Caton) Martin in Montevallo, Alabama. Growing up on a farm, things were sparse as the family had no vehicles and no electricity in the home. A round trip to a gristmill near Selma to grind corn into meal was an all-day affair. His family's first vehicle was a Model-T Ford.

Corporal Martin served on a Regimental Combat Team in the U.S. Army, Company H, 2d Battalion, 7th Infantry Regiment, 3rd Infantry Division, during the Korean War. He had enlisted in the Army on June 12, 1950, a mere 13 days prior to the outbreak of hostilities on the Korean Peninsula.

Martin's basic training lasted some 13-14 weeks at Fort Knox, Kentucky. At the end of his training, he was given 15 days leave en-route, then he boarded a troop train out of Chicago to a base near Seattle. They then flew out of Seattle to Tokyo via Anchorage and the Aleutians. From Tokyo, he went straight into the Korean War aboard a maritime vessel.

As the vessel approached Wonsan, North Korea, an LST pulled alongside, dropped the nets, and he and other soldiers climbed up the nets to get aboard. He actually landed with fellow soldiers at Wonsan, then moved north along the coast to join the fighting.

From 15 November until December 25th he took no baths, did not shave, and ate C-rations when he had them (some days he didn't). He lost his canteen in some of the fighting, so he crushed snow into liquid to have water to drink. When he lost his can opener in the chaos of fighting, he used his bayonet to open cans.

On December 19th, 1950, he fought with his machine gun all day covering for his unit as it moved. This went on until late afternoon. To get him out, the platoon leader had to bring up a tank to cover for Martin's movement. Had the North Koreans only moved to his rear, they would have killed him, but Martin's thankful they didn't see to do that.

For his actions that day, Martin received the Bronze Star with 'V' for valor. During the war, he also received a Purple Heart from being wounded. All in all, he received four major battle stars.

His basic pay was $70 per month, with another $45 per month added for being in combat. He had nowhere to spend the money, so he just sent it back to his family in Alabama.

By Christmas Eve, 1950 when an Allied retreat was nearing completion, he as a machine gunner was involved in fighting a rear-guard action in temperatures that were 20-30 degrees below zero, sometimes even colder.

After so many weeks of no bathing, no shaving, eating frozen C-Rations, fighting off frostbite, and being infested all over with body lice, Martin said he and his fellow soldiers "smelled like pigs" by the time they were finally taken aboard a ship at Pusan to allow him to recover from a severe case of laryngitis.

An Army quartermaster later set up a tent along the banks of a river, he was cleaned up and given clean clothing, and then was sent back into combat.

One compatriot and fellow soldier was his Squad Leader, Corporal "Mike" Hiroshi H. Miyamura (affectionately called "Hershey" by fellow soldiers), who was awarded the Medal of Honor (and later retired back to Gallup, N.M.).

Martin and Miyamura and Company H had fought nearly all night near Taejon-Ni on 24-25 April 1951 and were being overwhelmed by the enemy. Near daybreak, Martin and many of the company escaped. He assessed the situation and decided to crawl *toward* some of the enemy in front and find an avenue to get out, since their backsides were totally surrounded by North Koreans.

Unfortunately, Hershey was taken prisoner by the North Koreans after covering the company's movement, but not before he killed close to 50 North Koreans with machine guns and bayonets in hand-to-hand fighting when his bullets ran out.

Miyamura was ill-treated by his captors, as he told Martin during a get-together they had many years after the Korean War. They were especially displeased that he was of Asian descent, and they made him pay.

On one occasion, Martin's H Company heavy weapons outfit (commanded by a Lieutenant Smith from Maine) was ambushed in a rice paddy. A fellow soldier next to him was shot in the mouth, and Martin was also wounded.

Another soldier was assigned to assist him to the rear lines, and along the way they captured two North Koreans. Martin's escort seriously considered just shooting them, but Millard talked him out of it as the North Koreans pleaded for their lives. They tried to get an American tanker unit to accept the prisoners, but they didn't want them either.

He was eventually moved to a M.A.S.H. unit for treatment. When he had recovered enough to go back into action, he was handed a rifle—and not much else—and told to re-join his unit.

He had no idea where his unit was in action, so he started walking north. Along the way, he saw a festive commotion and stopped to watch a USO show being put on by stars of the Grand 'Ole Opry (he remembered Grandpa Jones was one of the performers).

Other than the horrendous cold and harsh combat conditions he experienced, Martin had another strange, chilling encounter while taking turns sharing a foxhole (on 6-hour shifts) with a 16-year-old fellow soldier—a Private Cooper—from West Virginia.

While alternating with Cooper, Martin jumped into the foxhole and tried to go to sleep, only to be awakened by a kicking, screaming body sliding down on top of him. It was an enemy Chinese soldier who was also trying to take cover in the same foxhole.

The Chinese soldier almost had a heart attack, and Martin had some severe palpitations and great angst before the Chinese hurriedly scampered out of the hole and ran zigzagging down the hill to get away.

Things happened so fast, Martin and Cooper were unable to draw a bead and shoot him, despite taking several shots at the fleeing soldier.

Martin recalled an eerie thing about fighting in Korea. As American soldiers fought back and forth, North Korean civilians would shadow their movements, even squatting and waiting for something to happen when movements were stalled. If a soldier discarded an empty cigarette pack or empty C-ration tin, a North Korean would run and pick it up. He finally realized the isolated North Koreans had never before seen white men.

After his time in the Korean War, Martin attained staff sergeant rank by the 18-month mark, which was fast, and was sent back to Fort Jackson. Then he went to NCO School and later a Weapons and Leadership School. He had stops at Fort Riley, Kansas as well as a base about 30 miles from Indianapolis.

When asked, he volunteered for a tour in Germany, despite having been back stateside less than the required minimum time to be able to qualify; a waiver was issued to allow him to go to Germany. He also became a first sergeant along the way.

Martin left the Army and later joined the Air Force. He pulled a tour in the Vietnam War and served until he could retire at Craig AFB, Alabama.

In the Air Force he had training at San Antonio and was sent to Amarillo to a mechanics school. Arriving there, he was told they needed drill instructors and would give him an extra stripe if he would take the job, which he did for some nine months.

He later had a tour at Perrin AFB, Texas, which had fighter interceptors. He was a crew chief and later a maintenance flight chief.

Then he was off to Craig AFB, in Selma, Alabama, where his Maintenance Office boss was a Lieutenant Holland, an Ole Miss graduate who knew very little about the minute details of the job. When Holland's boss had frequent briefings, the lieutenant always took along Martin for his expertise.

Holland once asked him how he managed to have a program so much better than others, and Martin replied "I do it like the Army would do it," a reflection of the many things he had learned along the way.

After retirement from the Air Force, he continued to serve the government as a civil servant for some eleven years. Martin now lives in Helena, Alabama.[1]

As a note of interest, a granddaughter of MOH recipient Hershey Miyamura's, Lieutenant Marisa Miyamura, was serving in the U.S. Air Force (as of this writing) at Scott Air Force Base, Illinois.

KURT MILLER

Kurt Miller was born in May, 1931, in Cleveland, Ohio. His father had owned a manufacturing company in Germany, but saw the light regarding changing world events and migrated to the United States after the end of World War I, settling in the Ohio area.

At age ten, his dad got him a job as a caddy on the golf course, and he won a lot of money against older players as his golf game improved. He was rambunctious and dropped out of Cleveland Heights High School in the 10th grade. He would later test for proficiency and ultimately enter and get a degree in Industrial Engineering from The Ohio State University, courtesy of the G.I. Bill.

At age 16, Kurt signed up for the Air Force, and entered active duty when he was only 17 years old. For the first two years, his job was working on and installing radios and electronics in Air Force aircraft.

When the Korean War broke out on June 25, 1950, Miller was doing his regular job at Misawa Airbase, Japan. The next day he was in Seoul, South Korea acting as a Forward Air Controller (FAC). He traveled in a jeep that had two radios, one for ground-to-ground communications and one for ground-to-air communications. He called in strikes to aircraft from personally observing a target, or getting information from another ground unit, then relaying that location to the attack aircraft in an effort to thwart the thrust southward of the North Koreans.

He started with the Army's 24th Infantry Division and then the 25th Division, but ultimately worked with several other countries' forces, to include Republic of Korea (ROK), British, Turkish, and others.

In my interview with Miller, he said that most times he was fighting right along the front lines. Temperatures often reached at or *below* zero. With wind chill effects, temperatures were well below zero. Frostbite was a big problem, and finding fuel for a fire to keep warm was almost impossible. They often took small amounts of gasoline from their jeeps or other vehicles to build a short-lived fire to get warm, if only for a few minutes.

Miller said casualties were high among FACs, something not unexpected when one considers they were usually on the front lines.

He and the American forces were pushed all the way down the Peninsula to Pusan, then he headed back up toward Seoul as American forces cut off the North Koreans after MacArthur's landing at Inchon that led to the American forces fighting all the way north to the Yalu River until the Chinese jumped in the fray. He said he was standing on bluffs overlooking the Yalu River when the Chinese came across.

Miller had a lot of respect for General Douglas MacArthur whose favorite saying was, "There is no substitute for victory." It was demoralizing for Miller when President Harry Truman fired MacArthur over their disagreement about pressing the fight over Korea on to a final, decisive victory. He felt like a lot of lost American lives went to waste over that decision. He had thought long and hard about becoming a fighter pilot, but stated he decided against it after that one decision that he felt was so misguided.

During the war, U.S. Marines and Army soldiers served six months then rotated home. Such was not the case for Miller. Miller fought continuously in Korea for almost 18 months. He said he was awarded five Bronze Stars and one Purple Heart.

When he came back stateside, he shipped a Russian revolver back for a keepsake, but it was later stolen from him here. He also sent another interesting item—a radio he removed from a Russian tank captured and disabled in the fighting. The fascinating thing about the radio was that all its tubes had been made in America (RCA brand!).

Miller was in the Air Force in Florida another six months before exiting the Air Force, then heading off to Ohio State for a degree. He mentioned that it only cost him $75 per quarter to attend Ohio State University back then, an astonishing figure considering today's tuitions at institutions of higher learning.

A retired American general taught Miller to fly, and he bought his first aircraft (a Stinson) when he had accumulated a mere ten hours of flying time. He later had a Commanche and accumulated over 5,000 hours in that aircraft; he has more than 9,000 hours total time now.

Kurt Miller held four real estate licenses in Ohio, he had a telephone business when with ITT, and he sold properties along Palm Coast, Florida, until the economy went south in 1969.

Along the way he was married to his first wife 1956 to 1963, and then a second wife from 1969 to 1978. He has two daughters who are both nurses.

He currently lives in Montgomery, Alabama and runs a landscape business (at the young age of 81, soon to be 82). Staying active keeps him young and alive (his words!).[1]

HAROLD O. CHITWOOD

"Chitwood broke off his attacks and pulled back to try to protect Joe McConnell"

(Photo courtesy of Harold Chitwood)

Harold Chitwood was born August 5, 1930, to Herman O. and Demile Chitwood in Gadsden, Alabama. The family lived in the country a few miles from town, and his parents taught school in a small three-room country school.

His father was a principal at the school and had been hired to teach before ever getting his college degree (by taking a test). They taught at a time when things were tough; Harold remembered his parents talking about once going six months without being paid. They later got their college degrees.

Wages weren't much back in those days, either. Harold himself recalls working for neighbors for 50 cents per day.

In 1943, the family moved to Boaz, Alabama, with Harold graduating from Boaz High School in 1947. He subsequently attended Snead State College and the University of Alabama, but decided later to go into the Air Force when he was only one quarter shy of his degree.

He was first at Perrin, Texas, for a month of Pre-Flight school, then headed to Spence Field in Moultrie, Georgia for six months of basic where he flew T-6s. That was followed by assignment to Bryan, Texas, where he flew T-28s, T-33s, and F-80s. In all his training, he got his commission at the same time he got his pilot wings. Chitwood was part of a class known as *52-Charlie*, that had 420 pilots graduate in May,

1952. Most flew combat in either Korea or Vietnam. One, General Jim McDivitt, was later an astronaut in the Gemini and Apollo programs.

After graduating, he was on to Nellis AFB, Nevada, for aerial combat training in F-80s and F-86 *Sabres* as they were phased into the inventory. There he learned all the combat maneuvering, tactics, and gunnery needed to survive in a combat environment.

The Korean War had started in 1950, and later on he and some 10 other pilots were aboard a C-54 with then-Captain Joe McConnell on their way over to Korea. McConnell was a veteran of World War II and would become a triple-ace. Chitwood was just a "shave tail" second lieutenant.

He was first assigned to the 51st Fighter Interceptor Wing, 39th Fighter Squadron, flying F-86s at K-13 (Suwon) in September, 1952. He would be in South Korea until near the end of June, 1953.

Another fighter outfit, the 4th Fighter Interceptor Squadron, was at K-14 (Kimpo) in Seoul. He recalled that Osan Airbase was under construction at the time (a base this author remembers well, having often flown C-130s in and out of Osan in the mid-1970s) .

Chitwood and the other squadron pilots fought all over both sides of the 38th parallel, often up near the Yalu River. One of his most memorable missions was on April 12, 1953, the day that McConnell, who became a top Ace with 16 MIG kills, got shot down. On departure, Chitwood was number 3 in McConnell's 4-ship of F-86s, and they had already split into two elements to made it easier to maneuver and dogfight.

A Russian MIG pilot and one of their top Aces, Semen Alexeivich Fedorets, first shot McConnell as McConnell was shooting another MIG. The Russian came in low underneath, and, despite McConnell's wingman's call for McConnell to break, managed to hit McConnell's Sabre.

Despite that, McConnell did a tight barrel roll and caused the Russian pilot to overshoot him, which allowed McConnell to then maneuver his damaged plane behind and shoot down the Russian Ace before limping off using less than 85 percent engine power, that was all that was available. He eventually bailed out over the Yellow Sea.

Chitwood broke off his attacks and pulled back to try to protect Joe McConnell, all the while radioing Chodo Island for Air-Sea Rescue. A rescue crew of Donald Crabb and Robert Sullivan on a helo of the 581st AR&C Squadron was in the area and maneuvered right to the area where

McConnell bailed out and picked him up almost immediately. Chitwood said McConnell was in the water only one or two minutes.[1]

A flight of four other enemy saw the smoke on McConnell's aircraft and came charging in to finish him off, but immediately fled north once Chitwood and the others in their flight began an attack against the enemy 4-ship.

After getting his 16th kill, McConnell was sent home and was flying experimental F-86H models at George AFB, CA. The H-model had bigger engines and other modifications. On one test mission sortie, McConnell lost primary flight controls. By manipulating flaps, speed brakes, and engine power, he was able to nurse the ailing aircraft back to the runway (extending the speed brake would raise the nose, and extending flaps would lower the nose, so an expert pilot could improvise and exert some degree of flight control in this type situation).

The next day, the very same thing occurred, but this time McConnell was not successful. He was too low to bail out and lost his life in the crash.[2]

Per Chitwood, his squadron in the first 10 months knocked down 65 MIGs without having a single American pilot killed (he stated they did lose some aircraft, but no pilots).

He also stated in his squadron was a flaming red-headed, aggressive major with a big mustache. The major was shot down and became a POW of the North Koreans at a time when they were accusing Americans of inflicting germ warfare by dropping infected bugs (a propaganda ploy).

During an interrogation, one of the red head's tormentors slapped a big bug on a table with that same accusation. The major promptly ate the bug to disprove their allegations, so they hospitalized him for several days until they were sure he was o.k. The major was released at the end of the war.

Another notable pilot with Chitwood was Harold Fischer who he said got credit for 10 aerial kills, but in actuality, per Chitwood, had 14 kills. Fischer was captured and taken to Manchuria as a POW. At the time of the Armistice, he was not repatriated, but he eventually did come home.

Chitwood described Fischer as a "Jeckyll and Hyde" type who was normally very quiet and unassuming, but turned into a vicious hunter/killer after strapping into the cockpit. He once pulled so many G's that his gun switch inadvertently was turned off. Fischer twice tried to ram the enemy MIG to take him down, until he realized the gun switch was off. He turned it back on, and promptly shot the MIG down.

Fischer's aggressiveness was manifested in other dramatic ways. He intentionally buzzed North Korean airfields (such as Antung and others), flying low over their alert shacks in an effort to lure them into scrambling airborne so he could shoot 'em down. When Fischer finally went down, it was only because he had shot up a MIG and his air intake ingested debris from the crippled MIG.

Chitwood said another 52-Charlie classmate, Quinn Fuller, was in a 4-ship when Fuller was shot down and surrounded by enemy. Three pilots strafed around his position as he tried to defend himself using his .45 cal pistol.

A rescue helicopter came in with a looped sling that a downed pilot was supposed to wrap around himself for rescue. Fuller hastily draped his elbow across the sling and held on until his rescuers could lift him up and get him out to safety. Fuller was so exhausted that he passed out once they set him down in safe territory.

Another pilot was in his squadron and flying a Sabre when he took 37-mm rounds in each wing, one round in the fuselage, and one at the flaps/aileron junction. He had 50 holes in his plane, as well as nine bits of shrapnel in his head and neck. Thompson's fuel tanks were ruptured, and he was on emergency flight controls with no radios and few instruments working. He was flying by the seat of his pants on the proverbial "needle, ball and airspeed," and above a 10,000 foot thick layer of clouds.

The pilot shut down the engine, dropped through the clouds, re-started the engine, landed and ran off the runway (both tires had been shot off!). He also passed out afterward, but lived to fight another day.

For a while, a U.S. Marine pilot, Major Jack Bolt, was a flight leader of Chitwood's. Bolt was already a fabled pilot, having flown *Corsairs* in WWII as part of Gregory "Pappy" Boyington's *Black Sheep Squadron*. In the Korean War, Bolt flew Marine F-9-F fighter aircraft, then requested to fly F-86s with the Air Force. He was assigned to Chitwood's 39th Fighter Squadron for several weeks and led many combat sorties. Bolt was also an Ace.

Another incident Chitwood recalled watching was the crash of Ted Williams, the *Splendid Splinter* of the Boston Red Sox, maybe the greatest hitter ever. He said Williams had 20-10 vision, and was flying an F-9-F for the Marines in Korea. Williams' plane was damaged and smoking as he was returning from combat for an emergency landing, accompanied by several Sabres of the 39th squadron who were trying to assist.

Williams had been advised to bail out, but chose to take his chances with a crash landing because the F-9-F cockpit was so small he feared he would damage his knees and not be able to return to major league baseball.

Chitwood and others saw the plane arriving and flying a long, wide pattern. They jumped in a vehicle and raced toward the end of the runway.

He remembered the plane burst into a big ball of flame as he neared touchdown, and all he could see was the aircraft nose and top of empennage in the fireball as the plane skidded to a stop. Fire trucks immediately doused the plane, and Williams returned to baseball with many more great all-star years in the American League.

Chitwood was not credited with a MIG kill, but he was credited with damaging three enemy aircraft. He left Korea in June, 1953, and was re-assigned as an instructor pilot at Nellis AFB where he served until leaving the Air Force in May, 1955.

While at Nellis, he learned about a North Korean MIG pilot who defected into K-14 and created quite a stir. Takeoffs and landings at K-14 were normally to the north. With the radar off for about 15 minutes that day, the MIG defector streaked in and landed to the south, meeting an F-86 on the runway that was attempting to take off to the north. Controllers were yelling over the radio at him trying to stop him, but he continued.

After rollout, the defector saw eight F-86s sitting to the side on alert, so he just taxied over and shut down amongst them. An American pilot jumped out and ran over to greet him, and an Air Policeman on a jeep pulled up to take the defector with him and place him in custody.

The defector eventually became a college professor at the University of Central Florida. Years later, when a MIG and an F-86 were dedicated at the Birmingham, Alabama Southern Museum of Flight, Chitwood met the defector, which was quite a treat.

After the Air Force, Chitwood finished his college degree at Alabama and moved back to Boaz in 1957 where he signed on with Gold Kist to be trained as a Plant Superintendent of a chicken processing plant. The first order of business was to go to Canton, Georgia, to learn how a chicken plant actually worked before returning to Boaz to supervise its plant.

At that time they were processing 240,000 chickens weekly. Chitwood eventually worked his way up to become CEO of Gold Kist in 1991, and they eventually processed 15,000,000 chickens weekly. Much of their output went to fast-food chains.

Within the company, he had over 22,000 employees (one of whom was my friend, Mike Bishop of Cairo, GA), and Gold Kist operated 100 farm stores in Alabama, Georgia, and Florida, plus a few in Texas. With Harold Chitwood at the helm, Gold Kist at one time was number 389 in the *Fortune 500* listing.

Along the way of his life's adventures, Chitwood married Fern Nash of Gadsden, and they raised three boys to adulthood. None of the three ever expressed any desire to be in the military, but all are successful.

He became involved with other civic things such as heading up the *Greater Atlanta Boy Scouts of America*, setting a terrific example of leadership for those young men.

Nearing his 83rd birthday at the time of this writing, Harold and Fern Chitwood enjoy the retired life in the Boaz/Albertville area. In the summers they sometimes travel to the Alabama Gulf Coast to relax in the sun, get with old friends, and enjoy a good restaurant.

Our nation is grateful to those courageous airmen such as Harold Chitwood who heroically served their country in war time, then achieved such notable business success while continuing to serve the American public in such a noble fashion.[3]

ROY H. JOHNSTON

"You see that hill over there, we're gonna take that hill tonight."

MSgt Roy Johnston Camp Evans, S. Vietnam (Aug '67 Photo)

Retired Marine Master Sergeant Roy H. Johnston was born April 16, 1928, to Felix and Lottie (Sasser) Johnston in what's known as the Burnout Community between Opp and Brantley, Alabama. He served in Korea and South Vietnam.

Johnston joined the U.S. Navy on January 8th, 1946, and was trained and served almost three years as an electrician on the submarine tender USS *Sperry*.

He decided he was "tired of such close quarters" as that aboard the *Sperry*, so he left the Navy and went back to a Troy high school to complete what was left of his high school education (he only needed one half-credit!).

With diploma in hand, Johnston joined the U.S. Marine Corps—he just wanted to find out what it was like to be among "a few good men," and it didn't take long once he reached Parris Island for his training.

He started Marine Boot Camp in the summer heat and ran into what he could only describe as "a hell hole." Things improved as he got into better physical shape, and he was able to better tolerate the demands placed on new recruits.

He remembered his drill instructors well. There was a Corporal Jordan, a Sergeant Seig (German descent), and a Staff Sergeant Kvietkus, all of whom he described as "sadistic S.O.B.s" which made them well-qualified to be Marine boot camp D.I.'s.

He remembered one D.I. in particular that always made them remove their caps at the most inopportune times so the mosquitoes that were abundant in the area "could feast on their shaved heads." He laughingly taunted the recruits in a loud voice, "Mosquitoes have to eat too, you know!"

Johnston recalled that there were 69 recruits on the second floor of the barracks, and they would have to run at break-neck speed when the D.I. yelled, "Outside, Now!" Before the call, the D.I. had locked the exit door on the bottom floor, so, when 69 recruits slammed against the first floor door, it actually lifted the door off its hinges!

Once during drills, when the D.I. called out "Right Face," he inadvertently turned left. The D.I. accosted him and asked "Where're you from, boy?" When he answered meekly "Alabama, Sir," the D.I. yelled "That figures."

By the time he was 22 years old in 1950, Johnston was fighting in the Korean War. His unit landed there in August, 1950, at what he referred to as "just out of bean patch" as part of the second wave that was referred to as the "First Replacement—Draft." His unit joined the 1st Provisional Marine Division which was under the command of a one-star general. When they added two more regiments, they became the 1st Marine Division.

They fought as part of what came to be known as the "Second Drive on Naktong," and he went across the peninsula as a Private with *Item Company* while he was a radio operator to a sergeant in the 5th Marine Regiment. Captain Harold G. Schier was the Commander of Item Company.

He said Schier's outfit was the same one that had raised the American Flag atop Mount Suribachi on Iwo Jima in WWII. Johnston would later have the privilege of serving with Schier in Birmingham on an assignment as a recruiter.

In mid-September, his outfit was pulled back to Pusan, loaded aboard ships, then swung around to be part of the Inchon Invasion under General Douglas MacArthur's brilliant plan to cut off the North Koreans while they were south of the 38th parallel. He was part of 5th Regiment, 1st Marine Division, during the Inchon Operation.

After being cut off, the North Koreans moved at full pace all the way back to the Yalu River, but Johnston himself didn't make it all the way to the Yalu.

There was one particular operation that really stood out more than any other during his tour of duty. Just before dark one day, his commander told him and his fellow Marines, "You see that hill over there, we're gonna take that hill tonight." Johnston remembered thinking, "This doesn't sound so smart," but, as good Marines do, they don't question their leaders, so they headed out after dark to take the hill.

As it turned out, the hill was packed with Chinese Army soldiers, so many that it didn't matter how many the Marines killed, they just kept coming. When all the blood-letting was over, Johnston's unit that started with 180 Marines had *only 20* U.S. survivors who came down off the mountain. Estimates of Chinese killed were several hundred, possibly in the thousands, but the U.S. casualties were enormous.

They later determined there were 10 Divisions of Chinese that had the Marines surrounded on all sides. Johnston said the U.S. Air Force saved the Marines when they air-dropped metal bridges by sections that allowed them to build an escape route.

In 1952 at Quantico, Virginia, Roy married Faye Henderson (of Calera, Alabama). They had two children, one girl and one boy. Their daughter was killed at age 23 in a terrible automobile accident, something Johnston described as, "A traumatic event from which parents never recover." Their son has made a career in law enforcement.

In his own career, Johnston pulled two peace-time tours on Okinawa. He went back into combat in August 1967 through September 1968 when he went in-country to Vietnam as an Operations Communications Chief, a radio operator, and duty in a message center, plus duties in radio maintenance.

He flew into DaNang, South Vietnam, then headed on up to assume duties as Operational Communications Chief, where he was part of the 4th Marine Regiment (which he said in WWII had been captured by the Japanese on the Bataan Peninsula).

The area was about ten miles south of Khe Sanh which was under siege for quite a period of time by the North Vietnamese. The U.S. Air

Force lost a lot of planes during the Khe Sanh operation and had to resort to drops of supplies and equipment using techniques such as Low-Altitude Parachute Extractions (LAPES) where the loads were yanked out of the aircraft by drag chutes without the aircrews stopping to offload. The aircraft just skimmed across the runway a few feet off the ground without actually touching down.

"...the Marines at Khe Sanh were encircled by North Vietnamese troops, and sufficient forces were not available to break through the garrison. Consequently, (in January, 1968) General Westmoreland decided to use air power to disrupt an anticipated attack by the two enemy divisions... Before the siege of Khe Sanh was broken at the end of March, Air Force, Marine Corps, and Navy aircraft flew some 24,000 tactical sorties against the forces surrounding the base. Flying 2,500 sorties by day and night, B-52s dropped almost 60,000 tons of bombs on trenches and artillery positions. Air Force transports landed 4,300 tons of supplies and 2,700 troops at the Khe Sanh airstrip, despite hostile mortar and artillery fire, and parachuted some 8,000 tons of cargo to the defenders."[1]

Some of the rapid re-supplying to Khe Sanh occurred with what was termed Container-Delivery-System (CDS) airdrops by aircrews such as those that flew the C-130 *Hercules*—an aircraft with which this author is intimately familiar, since I have close to 3,000 hours flying time in three different versions of the *Mighty Herk*.

Johnston was awarded the Navy Achievement Medal and authorized to wear the Combat "V" via a citation signed by Lt Gen H.W. Buse, Jr., USMC, that read, "During two regimental displacements, he ably supervised the orderly and expeditious relocation of his section with little disruption to its normal communications support operations. As a result... he gained the respect and admiration of all who observed him and contributed significantly to the accomplishment of his unit's mission."

Master Sergeant Roy Johnston retired from his beloved Marine Corps on December 31, 1968, then headed to Houston, Texas where he sold frozen foods to organizations that managed the Astrodome, Astro-World, etc.

He later became a supervisor with The Oil Well, Inc., a company that at one time was the largest independent dealership in the Southeast United States. He subsequently leased three different service stations himself, but at age 60, he "decided that his feet were always too sore to stand on asphalt all day" and went into permanent retirement.

Today, at age 85, MSgt (Ret.) Roy Johnston still gets around well and has a memory like a steel trap. He recalls his many adventures with the U.S. Marine Corps as if it all happened just yesterday.

He currently resides in Luverne, Alabama, only a few miles north of the rural home where he grew up as a youngster. The house on the old home place on a dirt road in the Burnout community is still standing, a tribute he says to the excellent carpentry, materials and building techniques that were used during its construction.[2]

Roy Johnston is very civic minded and is active with the Luverne Rotary Club and their many fund-raising activities that provide scholarships for deserving students. We salute him for his many accomplishments, and thank him for his dedicated service to our nation.

JOEL E. LAWRENCE

"Two Wars, Two Purple Hearts"

Joel Lawrence in Army uniform

Joel E. Lawrence was born near Florala, Alabama, November 4, 1913, to Robert "Bobby" and Sarah (Adams) Lawrence. A highly patriotic man, he served in two wars and was wounded in each. I'm proud to know he is in my family (distant cousin).

During World War II, his Enlisted Man's Identification Card (ETO—European Theater of Operations), listed him as Sergeant, home address Route 2, Andalusia, Alabama, weight 152 pounds, height 6 feet 1 inch—not a big man in stature, but one with a huge heart and fierce determination to fight for his country and the freedoms of others who were not so fortunate!

He was in the regular Army for 12 years, and served in the European Theater in World War II for 27 months. He sustained slight wounds in Belgium during fighting there and received a Purple Heart while with the 29th Infantry Regiment, Co. I.

During the Korean War, he shipped out of Japan for Korea on July 13, 1950, and in rank of Technical Sergeant was fighting when wounded in the knee. His local newspaper, *The Andalusia Star News*, carried his story on October 14, 1950 under a headline banner "Army Casualty Back in Tokyo," and reported he was one of the few men in his unit who escaped alive in the battle for Hongdong.

After being wounded, he was evacuated to a hospital in Tokyo on August 8th. His wife, Flora (Phillips) Lawrence, who lived on 216 Scherf Drive in Andalusia, received a telegram from the Secretary of the Army stating he had been slightly wounded in action on 7 Aug 1950. She then received a postcard from Joel himself, dated 8 Aug 1950, from the Army's 361st Station Hospital (former Do-Ai Memorial) in Tokyo, saying he was "o.k., except for the arthritis in knees."

After recovery, he was returned to limited service as chief enlisted supervisor in a Tokyo weapons factory. He wrote Flora that he was working from 12-14 hour shifts, with output of the factory being thousands of weapons daily.

The same *Star News* article related Lawrence's feelings about the Korean War through an article that had appeared in a Tokyo paper, *The Nippon News,* a clipping of which Joel mailed to Flora.The story quoted Lawrence as saying, "I never before got into anything like this Korea war." In reading other stories herein such as those of Millard Martin, Kurt Miller, and Roy Johnston, it would be easy to understand why he felt that way about the Korean War and things encountered there.

The conditions were tough, and chill temperatures often got down to 40 degrees below zero. Front line troops often got lice on their bodies, shivered in foxholes, went hungry, and engaged fierce fighters from both North Korea and China.

They deserve much more than just being casually mentioned as a soldier who fought in "America's Forgotten War." By 1956, Joel was an instructor/advisor at Midwestern University in Wichita Falls, Texas. He was one of four SFC's, along with LTC Donald Zeis, Major Lee Harmon, and Captain James Crawford in that capacity. He retired from the Army on 1 Oct 1959, at Fort Hood, Texas, S.O. No. 188, in rank of SFC (E-6).

In my mind, it's hard to conceive of someone who fought so fiercely and was wounded in both World War II and the Korean War being retired at only E-6.

Joel passed away in 1973 and is buried in the Peaceful Acres Cemetery, in Opp, Alabama.[1]

LT GEN CHARLES G. "CHICK" CLEVELAND

"Integrity First"

Lt Gen (Ret) Charles G. "Chick" Cleveland

Retired USAF Lieutenant General Charles G. "Chick" Cleveland was born in Hawaii in 1927 to Army Lieutenant Orestes and Katharine Cleveland III. His father was assigned to the 19th Infantry Regiment which had a legacy extending back to the Battle of Chickamauga in the American Civil War.

As a youngster, his family moved around quite often due to his father's ever-changing assignments. During a tour at Fort Benning that started in 1929, his father passed away unexpectedly, and his mother moved Chick and his older brother, John, and sister, Jane, to Woodstock, New York. When his mother's father began failing in health, she moved them in 1940 to Elsmere, a suburb of Albany, to be near her father so she could help take care of him. In discussing with me those who influenced him the most in his life, General Cleveland quickly stated it was his mother who had extraordinary strength of character and will to do what it took to take care of her family. He described his mother as, "A five-foot-tall 'Banty Rooster' with a backbone of steel."

Chick Cleveland was like many other young men of his day and age. He was quite the athlete at Albany Academy and participated in football and basketball. He still follows sports to this day, especially his beloved Boston Celtics basketball team. He also became enamored with aircraft, from about age ten forward.

An excellent student, he skipped an elementary school grade and completed high school at the prestigious Albany Academy. At age 17, with the help of his mother who was by that time active in Democrat Party politics, he managed to get an appointment to the U.S. Military Academy at West Point where his older brother John was just finishing.

He described West Point as "tough." He was no longer BMOC. Many students were veterans of WWII, and most all were older, some in their early 20's upon entering. They could be intimidating to such a youngster.

It was at the West Point chapel that he spotted a beautiful, charming, intelligent girl named Frances Riedel, from East Orange, New Jersey. She would become a big influence in his life and career, and was the perfect complement.

Graduating from West Point in 1949, he was one of 176 cadets of 574 total grads selected for aviation training. He headed out to Randolph AFB, Texas, for undergraduate pilot training where he flew the T-6 (some flights were to nearby Seguin, where he flew his first solo).

From Randolph, he was on to Phoenix, AZ, for pilot training at Williams AFB. He was an excellent student flying the F-80 *Shooting Star*, the Air Force's first operational jet fighter.

After pinning on his aviator wings with Class 50-E on September 15th, he and Frances were married and commenced on a life-long mission of service to our nation. They would be a magnificent team that would take them to places of leadership that would greatly benefit our country.

Cleveland finished high enough to be assigned a front-line fighter aircraft. He was first assigned to the 31st Strategic Fighter Wing flying F-84s at Turner AFB, near Albany, GA. There, in the 309th Fighter Squadron, he learned the tools of the trade by dueling in aerial dogfights with the best in the squadron, a fellow named Al Smiley, whom he eventually bested.[1]

When the Korean War broke out in June, 1950, Lieutenant Cleveland was assigned to fly the F-86 *Sabre* with the 334th Fighter Interceptor Squadron at Kimpo Airfield (K-14) near Seoul. The F-86 was heavily armed with six .50 caliber machine guns. Quite the skilled pilot, in one 8-week period, he shot down four Russian-built MIG-15s. All were confirmed kills. He had two other "probable kills."

One was on July 11 and the other on September 21, 1952, after he shot up the enemy MIG and it began falling while trailing smoke. He did not, however, claim it as a kill because, under the rules in play, to be a kill there had to be confirmation of a fire that wouldn't go out, or the attacking pilot had to see the enemy aircraft crash, or he had to observe the pilot eject from the aircraft.

His wingman was Lieutenant Don Pascoe who encouraged him to claim the damaged MIG, but Cleveland declined to do so in his mission de-briefing. A fifth confirmed kill would have meant Cleveland would be designated an Ace, but, being the man of integrity he was, he refused to claim a kill unless he was positive a kill criterion was met. He extended his combat tour twice, but it was not to be, or so he thought.

Returning to Turner AFB from his Korean War assignment, Cleveland was transferred to the 27th Fighter/Bomber Wing at Bergstrom AFB, Austin, TX.

He became a project officer for conversion to F-101 *Voodoo* aircraft, a McDonnell twin-engine fighter/bomber known affectionately as the "One-oh-wonder."

On December 12, 1957, he attained a speed of over 1212 mph in the F-101, which was actually 5 mph more than the official speed record claimed by another pilot a few days later.

It was during this assignment to fly the F-101 that he encountered a very close call. During a formation flight, the lead began a turn, and he began to cross over to the other side. When he applied aileron control pressure, the aircraft immediately completed two complete revolutions in something akin to an aileron roll. At the same time, his radio went out, so he could only communicate with the lead aircraft using visual signals. He was "sweating bullets."

He was able to straighten up the aircraft only by using rudder controls, but was confident he had enough controllability to get the aircraft safely on the ground. He flew into the pattern with his landing gear and flaps down and used rudders to wiggle the aircraft back and forth to alert the control tower to his predicament.

Things were going smoothly until about 50 feet off the runway when an unexpected gust of wind or updraft began to roll the aircraft. Due to his conditioning as a pilot, he suddenly applied aileron to counter the roll, and the aircraft immediately rolled the opposite direction.

Remembering his predicament, he then applied rudders to straighten the thing up and get it on the runway in a landing that wasn't pretty, but one from which he could safely walk away. His skill had saved many millions of dollars in an expensive fighter aircraft.

By February, 1959, Cleveland was assigned to RAF Station Woodbridge, in England, where he was a flight commander and operations officer in the 78th Tactical Fighter Squadron.

He later became commander of the 92nd Tactical Fighter Squadron where on August 10, 1962, he was the first pilot to log 1,000 flying hours in the F-101.

Later, he was sent to Wright-Patterson AFB, OH, where he was project officer for the F-111 *Aardvark*, the General Dynamics swept-wing aircraft entering the inventory that was primarily a bomber, but tasked with many roles. Arguably, the F-111 was the most controversial aircraft to enter the inventory. Some high-profile crashes during development and early usage didn't help the situation. Afterward, Cleveland was off to Vietnam where he served as an executive assistant under Army General William Westmoreland who ran the U.S. Military Assistance Command there. He had a very high opinion of General Westmoreland.

Cleveland had many more exciting assignments and promotions, including attending the Royal Air Force College of Air Warfare, a HQ 3rd Air Force tour, Vice Commander of the 20th Tactical Fighter Wing, and Commander of the 3535th Navigator Training Wing at Mather AFB, CA. In that position, he initiated the concept of having a pilot/navigator combination in 2-seat fighter aircraft which proved to be so successful as the combat environment became so much more complicated.

He subsequently had tours in Air Training Command (ATC), the Pentagon, and Randolph AFB again, where he was Air Training Command Vice-Commander.

It was during the tour in ATC that an unusual event occurred, one that this author remembers well since I was in ATC at Vance AFB at the time. Cleveland was Chief of Staff to Lt Gen George Simler, a very outgoing, gregarious, and dynamic leader of ATC who was being promoted to 4-star rank and assuming a new job as head of Military Airlift Command. Simler had flown two combat tours in Europe in WWII and was once shot down, evaded capture, and returned to the Allies.

Cleveland was at the base on that fateful Saturday, September 9, 1972, when General Simler departed in a T-38 *Talon* with instructor pilot Captain Gil Gillespie to fly to Scott AFB, Illinois, to assume his new command. Cleveland heard the loud boom of the crash and the sirens of the fire-trucks and rescue vehicles as they rushed to the scene of the accident. He headed to the area and was dismayed by the sight of the pilots' remains in the mangled metal.

At the same time, Lt Gen William V. McBride was driving toward Randolph to assume command of ATC, and was due with his wife, Kathryn, at the front gate within 15 minutes or so. Cleveland sped to the gate and met McBride, and it was at that time that he learned a lot about leadership and generalship from McBride who was a calm, unassuming, and strong general.

McBride's first question after being given the bad news was, "Have the families been notified?" They had not because the accident had been so recent. McBride called all the staff together at the commander's office and dealt with the situation in such a calm, professional manner that it always stuck with Cleveland.

To this day, he and McBride are still friends, and McBride went so far as to endorse a new book about General Cleveland's becoming an Ace in the Korean War.

General Cleveland's capstone assignment was heading up the Air University at Maxwell AFB in Montgomery, Alabama, a prestigious job that he and Frances both dearly loved and one at which they both excelled in serving the academic world and their local community and state. He retired in August, 1984, to enter the civilian world.

At retirement after 35 years of military service, he had flown 145 combat missions and had accumulated some 4,500 hours flying time. However, the retiring Lieutenant General Chick Cleveland wasn't through, by any stretch of the imagination.

In the civilian world, he served as Director of the United Way in Montgomery for seven years, and later was Commissioner of the Department of Human Relations for the State of Alabama until Governor Guy Hunt left office.

In 1992, he was honored at a "Gathering of Eagles" at Air Command and Staff College at the Air University. He was honored again in 2010.

One might think that would be the end of story, right? Not quite. Remember that "probable kill" of that MIG in Korea way back on September 21, 1952? Cleveland's wingman Don Pascoe certainly never forgot about it, and neither did one of his West Point classmates, Captain Dolph Overton, another American Ace in the Korean War who left the service after his stint in the Korean War.

The subject had once come up at an American Fighter Aces Association meeting, and Overton boldly told Cleveland that he would prove that "probable kill" was in fact a downed MIG that would make General Cleveland an Ace in the fighter pilot world's most exclusive fraternity (there were only 40 jet aces from the Korean War).

Overton went to work. He found out that after the end of the Cold War, in an era of *Perestroika* and *Glasnost*, the Russians had released a great many military records. Many of those records came to be located in our National Archives, and Overton went to work comparing Russian records of aircraft losses to the date/time/flight profile that Cleveland had flown on that fateful September day in 1952. He got copies of the info which was in Russian and arranged for an interpreter.

With other supporters and other fellow flying Aces such as retired Major General Frederick C. "Boots" Blesse accompanying, the group went in front of a special Corrections of Military Records review board and presented the Russian records of a lost MIG that day.

The board weighed the evidence and decided there was finally enough information to re-classify Cleveland's "probable kill" as a confirmed "kill," which gave him the requisite five downed enemy aircraft that placed him in that exclusive fraternity and designated him as America's 40th jet Ace of the Korean War.

General Cleveland's story was recently published in a book titled *Once A Fighter Pilot: The Story of Korean War Ace Lt. Gen. Charles G. "Chick" Cleveland.* Author Warren A. Trest's account of Cleveland's story is an amazing recall of the life of one of our country's great heroes.

General Cleveland confided that many things have to happen for a pilot to become an Ace. First, a pilot must have the ability. Secondly, one must have the proper training. Third, a pilot must have the opportunity. And, by that he meant there are many outstanding pilots who never serve in time of combat and never get the opportunity. Fourth, one must have a certain amount of luck!

He recently was briefed that the Fighter Weapons School may be closed because fuel is expensive. He expressed amazement at such a decision, saying, "It's expensive to fly, but it's more expensive if fighter pilots are not trained."

Integrity First is the first of three core principles of the U.S. Air Force. The other two are *Service Before Self* and *Excellence In All We Do.*

According to Cleveland, those core principles were instituted during the time of General Ron Fogleman's tenure as Chief of Staff of the Air Force, and he considers them a lasting legacy to General Fogleman.

General Cleveland's steadfast integrity over the decades in refusing to claim a kill on an enemy aircraft without adequate proof is one of the finest examples of that first principle for all the young airmen serving in today's Air Force. In fact, it should be a model for all Americans, period!

He and Frances have a son, Chris (the oldest, who lives in Virginia) and three daughters. Two of their daughters, Jane and Alice, are Special Education experts and live near San Diego. Their youngest, Susan, was a graduate of Montgomery Academy and Auburn University. She lives in the nearby area.[2]

All of America salutes General Charles "Chick" Cleveland for his long-standing service to our country. He epitomizes the American fighting spirit and is to be honored for his integrity, skill, and principled role in keeping our country free.

JIM, ART, DON, CHARLES, AND CLARENCE NEWBAUER

"War is hard and one never knows what is next"

FROM LEFT: Private Jim Newbauer in Army uniform; Arthur and Don Newbauer, Sr. chance meeting in Hawaii in WWII; One of the Newbauer brothers, next to B-29 "Windy City"[3]

James L. "Jim" Newbauer, Private, U.S. Army, was born to Mr. and Mrs. Arthur Newbauer of Greentown, Ohio, in 1931. He lived his entire life in Greentown prior to enlisting in the Army in 1948.

Jim was in a family of six boys and three girls. Five of the boys served in uniform, and all six would have, had it not been the case that one brother had severe heart problems.

Jim's older brother, Don Newbauer, Sr., was with the U.S. Third Marine Division for four years in WWII in the Pacific when U.S. Marines invaded places such as Guadalcanal, Bouganville, Hellzapoppin Ridge, Puruata, Empress Augusta Bay, Torokina, Guam and other landings.

Another brother, Arthur, was overseas with the Army for two years that same time. A fourth brother, TSgt Charles Newbauer, born in 1924, was a flight engineer. Charles was in service during WWII, the Korean War, and Vietnam War-era. A fifth brother, Clarence, served in Oklahoma after the Korean War.

While in Pensacola, Don, Sr. had met and married Blanche Lawrence from Florala, Alabama.

On November 24, 1950, Jim was in Korea and wrote a letter to his parents saying that his outfit, Company F, 27th Infantry Regiment, 25th Infantry Division, was "waiting to go into action and asked his parents not to worry about him."[1]

A few weeks prior, on 17 October, 1950, Don had written his brother, Jim, from Pensacola saying, "War is hard and one never knows what is next, sure hope you are among the first to leave that stinking hole over there." Don Newbauer knew all too well about war and its hardships from WWII.

Don related that he wasn't sure if he himself would be next to "go off" to the Korean War. He had reported to Naval Hospital on schedule in anticipation of being called for duty, but had a high temperature and was sent home. His temperature got so high, he had been, "Out of my mind and weak... but am now recovering."

In the meantime, the Marines had cancelled his orders because he had strep throat. Don closed by saying, "Jim, be good and take care of yourself. We are pulling for you and looking forward to the day we can see you once again...."

On 21 November, 1950, Don wrote Jim again, saying, "Dear Jim: Received your letter several days ago. Glad to hear you are still among the living and making the best of your surroundings. We are fine and think of you always." In another part, he said, "Am here sipping on a cup of coffee and wish you were here to have a cup with me."

He went on to talk about having gone to a championship football game of his son Sonnie's where they lost 12-6. They had gone to New Orleans a week prior and lost 8-7, despite being ahead by 7-2 until 55 seconds were left in the game.

Don continued, "Haven't heard any more of my going off since they classified me not physically qualified for active duty. Just wish this mess was all over. Just heard on the news where some outfit reached the Manchurian border. I wonder now if they intend to stop there, or do like they did (at) the 38th (Parallel)."

Don closed with, "Well, Jim, that is about it this time. You be good, Jim, and take care of yourself and write me whenever you can, as I think of you every day. Christmas will soon be here so will wish you a Merry Christmas and Happy New Year in case another letter doesn't reach you. In the meantime, keep your chin up as we are pulling for you. Love, Don, Blanche & Sonnie."

The two letters of Don Newbauer's never reached Jim Newbauer in Korea. Both were returned to Pensacola with large writing on the envelopes, to wit: "Deceased. Robert B. Gough, Sr., Capt. Infantry." Large rectangular stamped boxes on the envelopes contained the following: VERIFIED DECEASED, AG SEC., GHQ, FEC. E. C. GAULT, Colonel, AFC, Asst. Adj. Gen. (Don kept those returned letters and gave them to his son Don, Jr., who provided a copy to this author).

The following article appeared in a newspaper near the family in Ohio:

Private First Class James Newbauer, 19, was killed in action in Korea on Nov. 27, according to word received Saturday by his parents, Mr. and Mrs. A. R. Newbauer, of Greentown. The young soldier had been in the Army two years, enlisting when he was 17 years old. A LIFE resident of Greentown, he attended Greentown High School. He was with the 27th Infantry Regiment, Co. F, at the time of his death. He was the fourth one of the Newbauer family to see service in the Armed Forces.

The military service of Don and Arthur Newbauer was ALSO mentioned, as well as the service of a brother Charles A., with the Air Force at Bowling (Bolling sp.) Air Force Base, Washington, D.C. It went further to state "He leaves two other brothers, Paul and Clarence at home and three sisters, Mrs. Mary Craven (of Greentown), Ida and Sally at home."[2]

Like so many families of their era, Jim and the other Newbauer men distinguished themselves as some of thousands of *just ordinary citizens performing extraordinary service* for their country in its time of need.[4]

A grateful nation thanks the Newbauer brothers of Greentown, Ohio for their service and many sacrifices.

WILLIAM "BILL" DUDLEY, CARLTON, AND BOB TAYLOR

"When the Chinese overran his unit in May, 1952, a sniper killed his commanding officer, and he was given a battlefield commission"

William "Bill" Dudley Taylor and his brothers, Carlton and Bob, are the sons of James V. and Gladys (Benton) Taylor in South Alabama. Gladys' folks were from around the River Falls area, northwest of Andalusia. James' relatives were from Wiggins Crossroads area, a few miles east of Andalusia, AL. The family lived in the Horn Hill Community, about five miles southwest of Opp, as the three boys and a sister, Calene, grew up and attended Opp City Schools.

Carlton left high school in the 10th grade to serve in the U.S. Navy during the World War II years, and he continued in the Ready Reserve until 1950 at the time the Korean War began. He later finished his high schooling during a long and distinguished military career.

Dudley (or "Bill" as his military compatriots addressed him) was born Jan 16, 1933, and grew up driving a school bus, earning $1 per day for the Covington County School System. My older brothers and sisters easily recalled his days transporting them to and from a rural Covington County school on the big yellow bus with no air conditioning. They only knew him as Dudley, and so did the rest of the community.

As a youngster, he liked to trap Quail as a hobby. With the sudden passing of his father, the family needed some sort of financial means, so Bill joined the military to earn an income. In his senior year, at age 17 in 1950, he joined the 1st of the 117th Field Artillery in the National Guard unit in Andalusia, part of the famed 31st Dixie Division. After Mobilization, he finished his senior year.

With mobilization for war underway, and knowing that Bill would be leaving, Carlton left the Navy Reserves and joined the National Guard so as to go off with his brother.

Carlton ended up in Japan, but Bill ended up on the front lines of a hot shooting war in Korea. When the Chinese overran his unit in May, 1952, a sniper killed his commanding officer, and he was given a battlefield commission as an officer. Later on, he suffered a serious injury when hit in the back and was awarded a Purple Heart. Younger brother, Bob, remembered later seeing a scar on his brother's back about the size of a baseball from the wound.

He later was the CO of an Advanced Training Battalion at Fort Huachuca, Arizona, and, during the Vietnam War, was CO of the 3/13th Field Artillery in the 25th Infantry Division.

Bill married his first wife, Ruth Smith, whose father, Nathan Smith, was from near the Burnout Community a few miles north of where he grew up, and made plans to serve in the Army as a career officer.

He served two Cold War tours in Germany, as well as a combat tour in the Vietnam War, during which time the worst injury he incurred was a sprained ankle from playing volleyball.

However, in Vietnam, he was exposed to the deadly defoliant *Agent Orange,* the latent effects of which were not really well understood at the time. Suffering from the effects of the exposure later in life, he was eventually tracked by the Veterans Administration on the *Agent Orange List.*

He and Ruth had a daughter, Beverly, and a son, William "Butch," who eventually became a Parole Officer with the State of Alabama.

The wear and tear and stresses of a military career became too much, and Bill and Ruth eventually divorced. While he was with HQ 3rd Infantry at Wurzburg Army Base, Germany, he met Maureen Urey, a Southern California girl, who would become his second wife

They were married in 1965 and have a daughter, Dr. Leslie Taylor Haussen, as well as a son, Clark Benton Taylor, who also served in Army Artillery and also retired as a lieutenant colonel. Clark married Julia Studstill of Opp, Al.[1]

Bill attended the Army's Command and General Staff College and earned a Master's Degree from the University of Oklahoma. His son, Clark, also has a degree from OU.

After a tour at the Pentagon, Bill retired at the rank of lieutenant colonel and settled in Ft. Walton Beach, Florida, in the Cinco Bayou area. As a civilian, he worked as a closing broker for Sandestin for several years, then worked with the State of Florida Investigatory Services.

The effects of Agent Orange finally caught up with him, and he died December 29, 1993. He is buried in Barrancas National Cemetery in Pensacola, Florida.

Carlton continued with his Army career, eventually retiring at the rank of Chief Warrant Officer.

Younger brother Bob Taylor served four years in the U.S. Navy as a Guided Missile Technician. His time at sea was aboard the USS *Canberra*

(CAG-2) which was only the second ship in the Navy to have guided missiles.

With his mother ill and needing someone nearby for assistance, Bob left the Navy and came back home. He eventually got a job working in the carpet business. His first job paid $30 per week, not much by today's standards. However, things changed for the better as Bob eventually established his own carpet business. He has stores in Opp and Crestview, Florida, and a clientele who have learned that he is a businessman who delivers a quality product.

The sister of the family, Calene, married Reese Barnes, also a soldier who had served in World War II. Bob has a vivid memory from his youth of Barnes returning from World War II, arriving at their home in a taxicab and unloading his Army-issued Duffel bag with all he owned in that bag. Calene passed away some two years ago.

Not to be outdone, Calene's granddaughter, Captain Evette Calene Barnes, proudly continues the family tradition of service in the military. She is presently serving in the Army, and has already served two tours in Iraq.

A grateful nation salutes the amazing Taylor family for their long-standing service to our national defense.[2]

DOUGLAS CAMPBELL

"The Hydrogen bomb just fell out of the bomb bay some 10 miles away"

When Douglas Campbell enlisted in the Army in 1943, his world changed drastically. He had grown up in tiny Laurel Hill, in the state's Panhandle, about 15 miles north of Crestview, Florida. By his own admission, he had never been more than 200 miles from home, had never ridden a bus or train, and had never seen snow or mountains.

Campbell was born in 1924 and attended college at Alabama Polytechnic Institute a short while until the military draft got after him in 1943. With no transportation available, he hitch-hiked to what is now Auburn University, about 180 miles from home. His college attendance was short-lived.

At age 18, he joined up at Gulfport, MS and took six weeks of basic training there. In August he was sent to Houghton, MI for a 3-month Aviation Student Training Program at Michigan College of Mining and Technology where the snow was heavy and deep in November. From there he was put into the navigator training program and sent to Hondo, TX in December, 1943, where he was a cadet who received his commission as a second lieutenant at the same time he completed navigator training.

About 20 percent—some 40 out of 200—washed out of the program, but Campbell completed it in September, 1944, and was sent to Walla Walla, WA to become part of a 10-man aircrew flying B-24 *Liberator* bombers. Edgar Gibson and Bob Wolfe were his pilot and co-pilot, and Robert DeRosa was the bombardier. They flew about 15 sorties, roughly for 100 hours, and were pronounced ready.

Having never navigated over water, and his crew never even having flown over water, he and crew were given a B-24 with orders to fly 9,000 miles—all over water—to deliver the bomber to New Guinea, with stops at Hawaii, Christmas Island, Tarawa, Guatemala Canal, and Nadzab, New Guinea. During the briefing, they were told some crews left and were never heard from again, not very reassuring to Campbell.

After arriving New Guinea and flying a few orientation missions with an experienced pilot—to places such as Rabaul, Wewak, and Solomon Islands—his crew was assigned to the 307th Bomb Group, 370th Bomb Squadron, on Morotai Island in the Halmaharas. Flown there aboard a C-46, they landed and were still on the taxiway when the airfield came under attack from the Japanese. They had to shut down the engines and jump out and run for cover in the woods and a nearby ditch until the attack subsided.

The crew flew 35 combat missions that averaged 12 hours each, mainly attacking targets in the Philippines and Balikpapan oil refineries in Borneo which were heavily defended by enemy fighter planes and anti-aircraft artillery, which caused the loss of many American B-24s.

On one occasion, his B-24 crew and two others sank a Japanese Navy destroyer between Celebes and Borneo. They even once bombed Saigon (now Ho Chi Minh City in Vietnam). In Brunei Bay they strafed and bombed shipping at treetop level, an amazing sight for a big bomber.

These men were young, very young! The average crewmember on his airplane was just 20 years old. Campbell recalled in his memoirs, "We all matured fast. I think we went from 18-20 year olds, to 26-28 year olds in only two years. You grew up or you didn't live."[1]

On one mission attacking oil refineries, his aircraft sustained damage with about 100 holes from enemy fire. They lost a lot of fuel, but miraculously were able to land at an Australian base in some islands called Tawi Tawi. Campbell recalled one other mission in particular that was especially tough. On June 13, 1945, his crew and another on their left wing were bombing oil refineries when the wing aircraft was struck by many enemy rounds and suffered a great deal of damage. The wing crew was losing fuel rapidly from holes in the tanks, one engine was shut down, and they were losing altitude rapidly, more than 1,000 miles from home airfield.

Facing a choice of ditching in the ocean or bailing out, the crew opted to parachute out over water. Campbell's crew circled overhead, observed 10 good parachutes, and directed a PBY rescue seaplane to the site that picked up most of the crew. Despite their best efforts, the navigator, one gunner, and a cameraman perished.

Campbell later recalled, "A lot of things we did during WWII, that were scary as hell and dangerous, we thought, 'That's just part of war.' We didn't look on them as heroic or daring, but just what was expected of flight crews."[2]

With his 35 combat missions under his belt, he was flown to Manila which by that time had been recaptured by the Americans. He and about 1,500 others were put aboard a ship with orders to return Stateside.

As they left the harbor, his ship collided with a tanker ship, almost sinking in the process before the captain could turn it around and off-load the returnees. A week's delay ensued, and another ship eventually brought him back to Tacoma, WA on a zigzagging ocean voyage that took 28 days to complete.

WWII ended just as the ship arrived Tacoma. Campbell caught a train back to Crestview (which was a 4-day trip), opted to get out of the Army, and got on with his life. He did stay in the Inactive Reserves though, which would eventually mean more combat when the Korean War broke out.

Campbell married Nita Bryan in 1946, and they had a son, Douglas, born in 1948. They later had daughters, Rita and Janet, as well. He and Nita both worked in Atlanta, GA, between the two wars, and Campbell also attended night school, with their son at a baby sitter much of the time. He recalled that none of the three were very happy most of the time.

The Korean War began in June 1950, and Campbell and many other experienced aircrew members were recalled to active duty in September. He was first assigned to Kelly AFB in San Antonio as a first lieutenant navigator flying aboard C-97 transports hauling troops and cargo to Japan to support the war in Korea.

The C-97 was the fastest transport at the time (jets didn't come into the inventory until 1957). Their return flights from Japan were usually loaded with the critically wounded from the Korean War. Most times 2 doctors and 3-4 nurses were aboard flights that had 80-100 wounded that needed urgent care in the United States at places such as Fort Sam Houston hospital in San Antonio.

In 1951, Campbell applied for pilot training and was accepted. In September he was assigned to Hondo, TX for primary flight school where he flew the T-6 *Texan* for six months. Afterwards, he was sent to Big Springs AFB, TX where he trained in the more advanced North American-built T-28s that were a lead-in to flying jets. He trained in formation, acrobatics, cross countries, navigation, night flying, etc. Of a group of five fellow pilots with whom his family socialized, two were eventually killed in the Korean War.

The pilots graduated at Big Springs after flying about 50 hours in the T-28 and 100 hours in the T-33, which he described as a 2-seater F-80.

About 40 percent of the class washed out, either voluntarily or because they couldn't progress in the program.

In October, 1952 he went to Luke AFB in Phoenix, AZ where he learned combat formation, air-to-air combat (mostly against P-51s), dive bombing, skip bombing, strafing, etc. in the underpowered F-84B. At graduation, he was given a month off before heading to Korea's Taegu Airbase (designated K-2) to fly the best version of the fighter, the F-84G that had bigger engines than the trainer versions.

Their war missions were usually interdiction or close air support of ground troops. Interdiction missions were 10-200 miles behind front lines, attacking dams, truck convoys, trains, warehouses, troop staging areas, and supply areas such as fuel storage sites. Troop support missions were usually dive bombing on the front line, strafing, or low-level napalm drops.

Campbell recalled austere living conditions. They were quartered in open barracks, about 20 pilots to each barracks. They had a latrine/toilet and shower 100 feet away. The toilets were holes cut in boards, about 20 holes in the latrines. To Campbell, the winter of '52-'53 was the coldest he had ever experienced, with temps in the 20s and blowing wind that took the chill factor down to around zero.

In Korea, he was promoted to captain and was given responsibility as a flight commander over eight pilots. He was five years older and was the only one who had served in WWII, which was a factor. When he had 65 missions, the Korean War ended, and he returned home without having to fly the 100 minimum others had to have to leave.

He had one harrowing combat mission where his F-84G left tip-tank would not feed fuel. He was instructed to drop his tip tanks at the end of the mission, which he did, but the one that wouldn't feed fuel didn't drop off. His commander then instructed him to bail out rather than try to land. The bailout systems had about an 80 percent success rate, so he opted to land at a high rate of speed for controllability instead of parachuting out. On landing, he touched down at 170 mph and the unthinkable happened when the front of the left tip tank dropped to the runway, scraped along and caught fire. As he slowed through 150 mph, the tank broke loose and slid along beside the aircraft in a giant fireball, with fire trucks chasing the flaming tank and the F-84.

His commander wasn't happy, but he saved the plane and walked away alive. He learned from the experience, and "later on read of 5 or 6 pilots being killed trying to land T-33s or F-84s with one full wing tip tank and one empty tank. The aircraft usually 'cart-wheeled' into the heavy wing."[3]

Having spent about seven months in Korea, he headed back stateside, 29 years old with another war under his belt.

Another aircraft and another overseas assignment followed later. Campbell was assigned to fly the RB-57B reconnaissance aircraft, one that would get up to about 65,000 feet. On one hairy mission, he and a buddy pilot, "Reg" Petty, were in a two-ship assigned to photograph Russian submarine facilities in Albania. Campbell flew the camera plane, and Petty flew the radar plane that could pick up on radar any enemy intruder. Their route took them over West Germany, Eastern France, across Switzerland and the Alps, into Italy and down to the "boot," then Northeast to Albania.

Passing 60,000 feet altitude, when he began to overrun Petty, he pulled back the throttles—a little too fast—and both engines flamed out. He lost all pressurization, and his pressure suit immediately inflated, which made doing anything such as moving an arm or swiveling one's head extremely difficult. He knew he had to glide down to 35,000 feet or below to get an engine re-start. He also lost all flight instruments, and his canopy iced over.

He couldn't see, so Petty talked him down, telling him, "Turn left, turn right, level your wings, etc." which enabled him to safely glide below 35,000 feet, at which point he finally got the engines re-started and de-iced the canopy. They safely completed the mission, but not without a lot of angst on the part of Campbell. To top it off, four Russian MIG-21 interceptors came up to see if they could get to the flight, but they couldn't get that high.

He had other very interesting close calls, such as the time he took a flight from Rhein-Main, Germany, to Athens, Greece. At departure, the runway was firm, but had not been scraped of loose snow. On departure roll, snow kicked up into the wheel wells and iced over in-flight, so, on arrival at Athens, the nose wheel wouldn't lock down because of the icing problem back in Germany.

They skidded to a stop with no nose wheel, which caused some underbelly damage—an 18-inch square hole where the plane slid on the runway—but he and his passenger survived without injury. Just another day in the life and work of pilot-extraordinaire Douglas Campbell.

In 1957, while stationed at Turner AFB in Albany, GA, some of the crews flew six RB-57D's to Yokota, Japan for some six months, with a mission to photograph the Far Eastern part of Russia and North Korea, especially the Vladivostok area and all the military facilities in that vicinity. On return stateside, some pilots, Campbell among them, rode back in a C-124, a big 4-engine prop aircraft affectionately called "Old Shakey" by many.

Around midnight, he and another B-57 pilot, along with the C-124 pilot and instructor pilot got up a game of cards, leaving the C-124 co-pilot, navigator and engineer in the cockpit flying the lumbering aircraft on autopilot.

About midnight, headed for Midway Island, and more than 700 miles from Wake Island, the #3 engine caught fire. When the engine was shut down, the fire continued to burn. Only after the nearly-panicked crew put the big bird in a high-speed dive from 10,000 to 4,000 feet altitude did the airflow extinguish the fire, enabling them to land safely at Wake. Another close call for Campbell.

In the summer of 1962, he was transferred to Kirtland AFB in Albuquerque, NM, and assigned to the 58th Weather Reconnaissance Squadron flying B-57s. "Weather" squadron was just a cover name. This unit had nothing to do with weather, but for real missions flew the B-57s on Atomic and Hydrogen Bomb testing missions on atolls some 400 miles or so away from Johnson Island in the Pacific Ocean.

A tri-service organization called Joint Task Force 8 carried out the dropping and surveillance of the weapons. There were about 35 airplanes altogether, including two B-52s that actually dropped the bombs. Some twelve B-57s were equipped to sample the mushroom clouds at altitudes up to 65,000 feet.

The atolls had to be close enough to Johnson Island so the B-57 sampling pilots could land quickly and be extracted from the aircraft because it was so "hot" from the radiation that a human could not touch the aircraft exterior.

A forklift was actually used to remove the pilot, and remotely operated fire trucks with high-pressure nozzles sprayed the aircraft after landing for some time until cool down could be achieved. The aircraft would be left to sit on the ramp for about three days during which no person was allowed within 50 feet. After those three days, the aircraft was checked with a Geiger counter to see if it was safe to be serviced and flown on another air sampling mission.

Campbell recalled, "The pilot would be taken into a tent where he would remove his outer flying suit, which would be put in a metal drum to be burned. Then some pressure suit people would help the pilot remove the pressure suit. He would then get in a shower and bathe until a Geiger counter would show he was clean of any radiation on his outer skin. Sometimes, it would take three showers to get it all off. The people assisting the pilots were all dressed in protective clothes, including gloves, boots, and special head covers."[3]

The tests were run by Lawrence Livermore Labs of Los Angeles and the Los Alamos, NM, Labs. He recalled the heat was so intense from 12-14 miles away that often the paint on the dropping B-52 would be burned off, and sometimes the windows would be cracked from the sudden change in temperature, when it went from -50 degrees to +50 degrees in an instant from the blast heat. Campbell himself would be flying about 30 miles away and would enter the mushroom cloud usually 30 minutes after detonation.

Campbell recalled one occasion when the Hydrogen bomb was so large that it would not fit correctly in the bomb bay of the B-52, so they dug a hole, put the bomb in the hole, and towed the B-52 over the hole to load it. It was so large that the bomb bay doors would not close. It hung out below the bomber and was about 6 inches off the ground. It was so heavy, the B-52 had to take off with a very light load of fuel, then refuel from a tanker almost immediately after departure. That was just the beginning for this fiasco mission!

When all aircraft were in position, the commander gave the word to drop the bomb. In a moment, they heard the B-52 crew say "uh oh" over the radio. The Hydrogen bomb just fell out of the bomb bay some 10 miles away from the intended target; however, it wasn't armed, so it couldn't explode. It sank in 15,000 feet of water, and our Navy tried to find the bomb, to no avail (high tech devices such as the one that found the Titanic wreckage were not available at that time). The bomb being a one of a kind weapon, that ended that test.

While writing his memoirs, Campbell recalled that of the twenty pilots flying those sampling missions, three that he knew of had died of cancer. He himself was diagnosed with throat cancer at the time. The pilots spent about four months at Johnson Island, and no doubt their radiation exposure was far more than the average Air Force pilot who was never involved in Atomic Bomb or Hydrogen Bomb testing/sampling.[4]

At retirement on 23 July, 1978, Campbell had actually served in three wars: WWII, the Korean War, and the Cold War with the Soviet Union, and retired after reaching the rank of colonel. He had come a long way from those lean days in the 1920s and 1930s in tiny Laurel Hill, Florida, where things ease along at a slow pace. His contributions were amazing.

After his Air Force days, he owned and operated a building washing business. Probably he was most proud of his volunteer mission work with the Mormon Church—a passion he pursued, and one which took him to places such as New Guinea, Australia, and other remote locations.

Douglas Campbell passed away in 2006, and is buried in Fort Walton Beach, FL. Many of his relatives still live in the Panhandle area. One of his brother Earl's grandsons is in the Air Force and flies C-5 cargo aircraft, carrying on that family tradition of military service to the nation. He has two nephews, Wayne and David, who are medical doctors in Crestview, Florida.

Douglas Campbell is gone, but his legacy lives on. One who grew up as just an ordinary man from a small town in Florida, but then lived an extraordinary life and made immensely valuable contributions to our nation's national defense. His story is one that young people everywhere can read and hope to emulate.

LUTHER H. STORY

"When last seen, he was firing every weapon available"

PFC Luther H. Story[1]

Private First Class Luther H. Story was born July 20, 1931, in the little Georgia town of Buena Vista, about 20 or so miles east of Columbus, home of the Army's Fort Benning where so many American Infantry soldiers and heroes have trained over the years.

Anyone visiting the Andersonville Civil War National Park Museum and Cemetery (as my wife, JoAnn, and I were on March 16, 2013), is likely to pass along Georgia Highway 26, if they're traveling through Columbus on the way to the National Park. Five miles south of Buena Vista is a small creek crossing the highway. The only thing that distinguishes that creek is the large brown marker on each approach end.

PFC Story was assigned to Company A, 9th Infantry Regiment, 2nd Infantry Division, during the Korean War.

Near Agok, Korea, temperatures were already very cold on September 1, 1950, when Story distinguished himself in a manner of sacrifice that only an American hero could.

His citation as published with General Order Number 70 on August 2, 1951, to accompany the *posthumous* awarding of the Medal of Honor speaks for itself:

G.O. No.: 70, August 2, 1951

Citation:

Pfc. Story distinguished himself by conspicuous gallantry and intrepidity above and beyond the call of duty in action. Savage daylight attack by elements of three enemy divisions penetrated the thinly held lines of the 9th Infantry. Company A beat off several banzai attacks, but was bypassed and in danger of being cut off and surrounded. Pfc. Story, a weapons squad leader, was heavily engaged in stopping the early attacks and had just moved his squad to a position overlooking the Naktong River when he observed a large group of the enemy crossing the river to attack Company A. Seizing a machine gun from his wounded gunner, he placed deadly fire on the hostile column, killing or wounding an estimated 100 enemy soldiers. Facing certain encirclement, the company commander ordered a withdrawal. During the move, Pfc. Story noticed the approach of an enemy truck loaded with troops and towing an ammunition trailer. Alerting his comrades to take cover, he fearlessly stood in the middle of the road, throwing grenades into the truck. Out of grenades, he crawled to his squad, gathered up additional grenades, and again attacked the vehicle. During the withdrawal, the company was attacked by such superior numbers that it was forced to deploy in a rice field. Pfc. Story was wounded in this action, but, disregarding his wounds, rallied the men about him and repelled the attack. Realizing that his wounds would hamper his comrades, he refused to retire to the next position but remained to cover the company's withdrawal. When last seen, he was firing every weapon available and fighting off another hostile assault. Private Story's extraordinary heroism, aggressive leadership, and supreme devotion to duty reflect the highest credit upon himself and were in keeping with the esteemed traditions of the military service.[2]

Private First Class Luther H. story is buried in the Andersonville National Cemetery, Section MA-6, Grave 9.[3]

Bridge marker on GA Hwy 26 near Buena Vista, GA (personal photo)

VIETNAM WAR

The Vietnam War was without a doubt the most socially-divisive war in American history. What began as a small-scale effort by American administrations (35 advisers in 1950) to assist South Vietnam in a counterinsurgency, ultimately turned into a quagmire.

In his inaugural speech in January, 1961, President John Kennedy had included the following regarding our national will and our opposition to tyranny everywhere: "Let every nation know, whether it wishes us well or ill, that we shall pay any price, bear any burden, meet any hardship, support any friend, oppose any foe, in order to assure the survival and the success of liberty. This much we pledge—and more."

President Truman on 27 June 1950 had included monetary assistance to France in Southeast Asia as they fought the communist forces of Ho Chi Minh and General Vo Nguyen Giap.

By 1954, the United States was footing three-quarters of the tab. President Eisenhower had first authorized military assistance programs, including the sending of military advisers to that country.

President Kennedy later expanded that commitment, including dispatch of the American Special Forces (Green Berets). By the end of 1962, the American commitment had deepened and more than 11,000 advisers were "in country..."[1]

Corruption was rampant in South Vietnam's government, Buddhists began protests, and South Vietnam's Chief of State Ngo Dinh Diem cracked down on the Buddhists. American dissatisfaction with Diem grew.

A critical event occurred when a Buddhist monk, Thich Quang Duc, self-immolated after abetting protesters doused him with gas. The protestors made a public spectacle, even blocking fire trucks that tried to extinguish the fire and end Duc's ten minutes of agonizing death. Other monks followed suit as protests grew and threatened to get out of hand.

JFK knew then that Diem had to go, and it wasn't long until a military coup killed Diem. As a result, control of the country reverted to the military and changed often, as if it were a game of musical chairs. In 1964 alone, seven different governments were in power in Saigon.

Eventually the war became a quagmire, with a peak commitment by the U.S. of 542,000 troops by 1969, in an effort that claimed over 58,000 American lives (including three from my hometown, Steve Stanley, Max Kersey, and Jimmy Darby). The war caused great social upheaval and resulted in massive anti-war demonstrations by crowds chanting "Hell no, we won't go."

By April, 1975, when my own 345th Tactical Airlift Squadron C-130 aircrews were launching out of U-Tapao, Thailand, to fly to Tan Son Nhut airport and evacuate refugees ahead of the Communist drive into Saigon, those words of President Kennedy would seem like only a distant, faint echo.

In the midst of the Communist onslaught, there was little teeth or meaning in Kennedy's words after our two decades of trying to help the South Vietnamese win the war and preserve their freedom and independence.

Despite the unfavorable outcome, it wasn't for lack of effort on the part of thousands of servicemen, as you will come to know from reading the following about so many of those gallant military members who fought so hard to assure freedom for the South Vietnamese.

Their stories remind us that, "War is hell," whether fought in France or Germany, on some South Pacific island, on a frozen tundra in Alaska, or in the jungles of Vietnam, Laos, or Cambodia.

The end result was the same, with blood, sweat, tears, loneliness, crying, dying, injuries, recoveries in hospitals, funerals, and memorial services.

Perhaps the worst of all was the angst of those family members back home dreading the arrival of a military sedan, followed by a knock on the front door by two soldiers and a chaplain in military dress uniforms with news that can never be good. Some things about wars never change.

The Vietnam War was no different in that respect.

SIR FRANCIS MCGOWIN

"He was knighted by Queen Elizabeth II for service to the United Kingdom"

McGowin, outfitted with parachutes, preparing for one of 2,147 jumps

Sir Francis McGowin was born in July, 1940, in Greenville, Alabama, into one of the most prominent families in South Alabama.

A great-grandson of Corporal James T. McGowin (Company D, 16th Alabama Infantry, Confederate States of America), Francis is one of the McGowin family who owned/worked in the Chapman Plywood Mills near Georgiana. He attended school and worked with the company until age 17 when he joined the United States Marine Corps on March 31, 1958.

His first five years were with the 47th Rifle Company reserves in Montgomery. Counting reserve and active duty time, his career in the Marines lasted some 27 years.

After going on active duty status, he eventually achieved the rank of master sergeant, but often occupied billets equivalent to Army colonel, with equivalent responsibilities.

McGowin worked with what was known as MAGIIS—Marine Air-Ground Integrated Intelligence System—and was involved in "getting computers to talk to computers" while with the Fleet Intelligence Center, Europe.

He at one time worked on developing war plans for Europe, and worked on many other projects at the highest levels, some of them highly classified. He said he saw the initial plan three months before the 1970 *Son Tay Raid* that went deep into North Vietnam—23 miles from Hanoi—to attempt rescue of American POWs.

McGowin said he lived and worked in some 51 countries and served three tours in Vietnam. He was wounded three times on his second Vietnam tour while with "L" Company, 3d Battalion, 3d Marines, which was Colonel John Ripley's outfit known as "Ripley's Raiders." Ripley was one of the most distinguished Marines in American military history. McGowin also spent nearly one and one half years in South Korea at the First ROK Marines Division near Pusan.

On his third tour in Vietnam, he was an instructor for Air Force pilots, teaching them Survival, Evasion, Resistance and Escape (SERE) techniques in case they crash landed or were shot down and faced capture by the Viet Cong or North Vietnamese regulars.

At one time in his illustrious career, he served with the French Foreign Legion and also the British Royal Marines. He was knighted by Queen Elizabeth II for his service to the United Kingdom, a status not achieved everyday by somebody from South Alabama (author knows of no others).

McGowin's service as a Marine totaled 27 years (5 reserve, 22 active duty). In that time, he graduated from Army Ranger School and made a total of 2,147 parachute jumps. Among his other training, he made some 1,000 SCUBA-dives.

After retiring from the Marines on July 29, 1979, he attended Lurleen B. Wallace Community College, then worked in the Forest Service for three years, to include teaching at the Dixie Forestry Department.

Afterwards, he worked another 18 years with the Veterans Administration covering Covington, Conecuh, and Crenshaw Counties, where he worked on behalf of military veterans, helping them obtain the medical benefits they had earned through their many years of faithful service to our nation.

Sir Francis McGowin is a coveted guest speaker and lecturer on many topics about historical subjects in general, and especially the War Between the States, in particular. He especially enjoys working with young people, having worked with the Boy Scouts many, many years, as well as going into schools for special presentations.[1]

McGowin is married to Virginia Ann Cook of Andalusia. Their children include a son, Steve, who spent 21 years in the Marines. Colonel John Ripley was Steve's regimental commander and promoted Steve to sergeant. When Steve deployed to Operation DESERT STORM, John Ripley's son was in the same unit with Steve. After the military, Steve at present works with Covington County's Sheriff Department.

Sir Francis and Virginia also had a son, Dan, another son, John (deceased), a daughter, Katrina (deceased), and a third son, Kenny.

DURWOOD LAMAR GOODEN

"Helicopter blades cannot out-duel enemy anti-aircraft fire"

Lamar Gooden during interview from Opp, Alabama B&B Restaurant

Durwood Lamar Gooden was born to parents D. B. and Sara (Reynolds) Gooden of the Beulah Community near Opp, Alabama. Lamar attended grade school at Beulah, then graduated from Opp High School in 1963. Typical music of his teen years included songs such as *At The Hop*, sung by Danny And The Juniors (1958).

Lamar attended Troy State College, graduating in 1967 with degrees in Business Administration and Marketing. During his junior year, he married his sweetheart, Lynda Bryant, also of Opp.

Following graduation, he headed off to the Firestone Tire and Rubber Co. Management Training Program, but while in Pensacola with the program, he received draft notices from Uncle Sam, as the Vietnam War was going hot and heavy.

Lamar signed up for Army Officer Candidate School, but first went through Basic Training at Fort Dix, New Jersey. From there he was off to Fort Benning, Georgia, where he received his commission as a second lieutenant on November 6, 1968.

It was generally understood that second lieutenants "knew little to nothing." They were disparagingly referred to as "butter bars" or "90-day wonders," or whatever other derogatory term could be conjured up by those who had been in the Army for a while (and knew *everything!*).

Before he ever considered aviation, he was in Infantry where he had an epiphany one day while training in Georgia swamp water up to his nose while wading and holding his rifle over his head. He heard a noise, looked up and saw a helicopter flying overhead and thought to himself, "That has to be a whole lot better than what I'm doing."

So, the next morning when the troops were asked if anyone wanted to go to flight school, Lamar was the first to volunteer, going against that old military service admonition to *never volunteer for anything*.

He and Lynda lived in Opp, and Lamar commuted each day to Fort Rucker, where he was first assigned duty as a headquarters training officer.

The day to leave for flight school finally arrived, and in the early part of 1969, he and Lynda loaded up all the worldly possessions they owned into a Pontiac Lemans and headed west to Fort Walters in Mineral Springs, Texas.

There he started flight training in Class 69-30 where he first trained in the OH-23D (1952 models with wooden rotor blades).

That training was followed by further training at Fort Rucker and Instrument Flight School, flying the TH-13 and then transitioning to the UH-1 Huey.

Lamar's class was the first that did *not* deploy to Vietnam immediately after completing flight training. Instead, he was sent to Fort McClellan, Alabama, where he was one of only two lieutenants there who had not yet pulled a tour in Vietnam.

On August 31, 1970, he was sent to Vietnam as part of the 1st Aviation Brigade, 13th Battalion, 162d Assault Helicopter Company. His station was Canto Army Airfield, about 80-90 miles from Saigon. He flew a lot, easily maxing out his allotted time the first month there.

Canto was next door to Bien Thieu Air Force Base, and his company had three platoons—the first and second used for troop combat insertions and extractions, and the third platoon helicopters were outfitted as gunships.

The routine was usually one day flying combat assaults, and the next day flying "ash and trash," words used to describe missions to ferry colonels and others around for routine meetings, etc. Their area of operations included from Saigon all the way down to the tip of the Delta.

With experience gained, Lamar eventually flew occasional missions over to Phnom Penh, Cambodia. They just took off about 0200 hours in the morning, in a flight of three, and followed the river. It was standard practice that one of the helos and crew never landed, staying aloft to provide safe cover for those on the ground.

The U.S. Air Force provided flight following. As they reached the border, the helo pilots reported over the radio "United States of America 1-2-3, at the fence at this time." Upon arrival, the crews and personnel being transported were given rides in expensive limousines, something to which sweaty Army helicopter crews were not accustomed. After a long day to and from Cambodia, they usually arrived back at home base in Vietnam at about 2200 hours.

Lamar's helicopter was shot down in March, 1971, with serious consequences. A sister unit had lost some folks, and Lamar and his flight of five helicopters were trying to recover those other troops, as some had been seen waving at flights passing overhead.

The mission was on a Monday, in mid-March, 1971, about 1300-1400 hours in the afternoon, and he had a load of 10 ARVN (Army of South Vietnam) combat troops aboard.

As he led his flight on an approach, at about 800 feet altitude, his helicopter began taking heavy fire from an enemy .51 caliber anti-aircraft gun with incendiary (exploding) rounds, and he could hear the distinct sound of bullets on metal.

He remembered a buddy flying on *Chalk 2* saying "Oh my God" over the radio as Lamar's helicopter's fuel tank was struck and exploded in a ball of fire. All the smoke detectors went off and cockpit alarms began screaming, then the helicopter slammed into the ground.

Lamar just remembered the horrendous impact and that his pilot's seat had slammed into the rear of the cargo compartment.

Despite his injuries, he unhooked his shoulder harness and seat belt, and rolled out of the mangled wreckage. As he rolled out, he dragged one badly injured (but still alive) ARVN soldier from the burning wreckage.

A good friend, Ken Bernard, flying in *Chalk 2*, also experienced enemy fire with his helicopter being shot down as well.

Ken managed to get about 50 feet from the wreckage when his exploded. They were both, along with another buddy in another chalk, all med-evaced to Bien Thieu hospital—Lamar just remembered stretchers/gurneys coming out to pick everybody up for transport to the hospital.

Waking up in the hospital, he pulled the covers back to look at his legs, and both were completely black from the incredible bruising from the crash.

Lamar never told his wife Lynda about the crash, but his mother back home became suspicious something bad had happened when she received a letter from him written on Red Cross stationery. The crash of Gooden's helicopter claimed a total of 9 lives that day. He was very fortunate to be a survivor.

After some 30 days of recovery, when Lamar got well enough to fly again, he went up with an instructor on a check-ride. When they began making an approach to landing, on the first three occasions he leveled off at about 200 feet altitude, an obvious psychological reaction to the severe consequences of having been shot down. It took some doing, but he ultimately overcame that problem.

Experiencing back problems, X-rays soon showed calcium deposits forming around the fractures to his spine from the crash. He was sent back stateside on August 31, 1971, one year to the day from his having been sent to Vietnam.

Gooden was assigned to Fort Benning, Georgia, in the Office of Management and Budget (OMB) office with a colonel, two lieutenant colonels, two majors, and six captains. In his words, "The job was so boring that the colonel answered his own phone." Lamar worked in the OMB office until February 28, 1972.

After discharge from the Army, Lamar returned home and went into business with his father-in-law as homebuilders. He later worked with his son, Brad, in a recycling business.[1]

Lamar Gooden considers himself a lucky man to have survived the war in Vietnam, especially considering what might have been on that unlucky day in March, 1971, when it was proven that helicopter blades cannot out-duel enemy anti-aircraft fire.

GENE AUTRY CONNOR

"The ground lit up like a Chinese New Year's celebration"

Connor at Eglin AFB 1961 (Personal Photo)

Major Gene Autry Connor, USAF (Retired), a native of Troy, Alabama, was born September 7, 1937. His mother's favorite singer was Gene Autry, hence the name. He graduated from Troy High School in 1955, then completed two years of higher education at nearby Troy State College.

Connor attended Air Force Basic Training at Lackland AFB, Texas, near San Antonio. From there he headed on to Harlingen, Texas for flight school in T-29 aircraft.

His basic and Aviation Cadet school lasted about a year, with his simultaneously receiving a commission as a second lieutenant and getting Navigator wings, then he was on to Mather AFB, California, where he did advanced Navigator training.

From there he was off to Stead AFB, Nevada, for Survival School where he learned to navigate through rugged terrain with only a compass, as well as survive off the land by eating such delicious morsels as snakes, rabbits, and porcupine.

His B-52 bomber training to check out as a Navigator was at a base in New Mexico. While en-route to Walker AFB at Roswell, Gene and Betty Carswell (also of Troy) were married. The training at Walker was followed by an assignment to Eglin AFB, Florida, for almost five years.

Their next door neighbor at Eglin was an Army Green Beret instructor who introduced Gene and Betty to the finer techniques of surviving off the land. The Green Beret would occasionally bring over alligator meat or plop down on his porch a croaker sack of snakes to be eaten. This always freaked out Betty, as she was not particularly fond of snakes.

Gene was at Eglin during a 1962 visit by President John F. Kennedy and Vice President Lyndon B. Johnson. As part of the festivities, demonstrations for those leaders included one of the infamous SAC scramble-alert launches.

Connor and crew were in the alert shack when the claxon rang. They sprinted for the aircraft, and the aircrews launched 8 consecutive B-52s with an average interval of 11 seconds between the liftoffs of all aircraft.

On takeoff, his aircraft got caught up in the wingtip vortices of those in front and rolled almost 90 degrees to the left before the pilots could correct the condition. It was a very close call, within a whisker of a deadly crash.

Per Connor, in a cost-cutting move, SECDEF Robert McNamara had Connor's unit moved to Barksdale AFB near Bossier City, Louisiana. His unit at the time was flying G-model B-52s, and the aircraft at Barksdale were the D model, so it was hard to see what savings were there to be realized.

After the Barksdale tour, he experienced an individual transfer to a B-52H squadron at Minot AFB, North Dakota. It was from Minot that Connor's unit forward deployed elements to either Andersen AFB, Guam; U-Tapao, Thailand; or Kadena, Okinawa for employment in the Vietnam War. By this time, Connor was qualified as both a Navigator and a Radar Navigator.

He watched some bomb drops through the "optics" (which he described as "basically the old Norden Bombsight—with slight modification") and could observe anti-aircraft artillery (AAA) rounds being fired toward his aircraft. His aircraft were shot at often over Vietnam, but luckily never hit. On one occasion, an F-4 flying MIG-CAP to protect the B-52s was hit by enemy AAA. The F-4 pilot had to bail out, but was quickly rescued by an American helicopter crew.

Connor recalled that because it took so long to affirm the intelligence and to get clearance to drop their munitions, the Viet Cong were often long gone by the time the B-52 strikes began, so there was not a lot of evidence of their bombing effectiveness.

The aircrews came up with the slang term "monkey manglers" to describe the imagined (and very real) damage done to jungle critters by their bombing runs.

The crews made their bombing runs from altitudes as low as 31,000 feet and occasionally as high as 43,000. There was not much to observe from those heights, but, on one particular mission where they were bombing suspected ammunition caches, the results were spectacular.

His crew was the center aircraft flying abreast in a 3-ship formation dropping 108 bombs from each aircraft. While counteracting enemy "95s" (95-mm anti-aircraft guns), they began dropping the bombs in a line and "walked" the bombs into the suspected site, then turned away and began their departure, all the while thinking not much was there, or perhaps they had missed the target.

Then, after a period of time on their exit heading about 60 miles away from the target, a mountain-side lit up like a Chinese New Year's celebration, with huge explosions that shook the jungles and lit up the skies as well.

Bingo! They knew then they had hit a mother-lode of stored ammunition, every bombing crew's dream!

Connor described some of the first B-52s he flew as having "*analog* computers with clear panels through which one could actually watch the component wheels turning inside." Components such as those would be considered "stone-age technology" by today's crewmembers, considering all the modifications and upgrades to the venerable B-52 over the years.

During one eventful R & R trip to Saigon (they caught a flight on a C-97 into country), Gene and other bomber crewmembers were flown on Army CH-47 helicopters of a 101st Airborne Division unit – sort of a "joy ride." On the day he flew, his helicopter took a hit from enemy small arms fire, which created more angst than joy among the joy riders.

The Army had been extra careful not to identify these Air Force passengers as B-52 crewmembers because word was out on the street that the North Vietnamese had put a price on their heads in order to get a crewmember in their hands. Their deadly bombing raids were termed "walking thunder" by the North Vietnamese because of the terror sowed by a string of dropped bombs coming out of the clear blue from the extreme high altitudes flown by the B-52s.

On one sortie out of U-Tapao, the aircraft he was aboard blew an engine shortly after takeoff (keep in mind that B-52s have *eight* engines). The crew chief happened to be sitting in the Instructor Pilot's seat; when the co-pilot looked over at the crew chief, he described the crew chief's eyes as being "as big as saucers" from his shock and surprise over all the emergency lights that lit up on the cockpit eyebrow panel, as well as all the noise from the alarms going off in the aircraft.

The crew circled over the Gulf of Siam for a while to get the problem taken care of and burn off fuel to come back for an uneventful landing.

During his time aboard the B-52, Gene Connor flew 155 combat missions over Vietnam and Laos, quite a record of service!

Major Gene Connor served his country well for 25 and one-half years (22 in the Air Force and 3.5 years prior with the Army National Guard in the 933rd Field Artillery unit at Troy, Alabama, where he worked all seven gun positions while assigned to that unit).

Also in his career was a stint as a Life Support Officer at Kunsan Air Base, South Korea. Another stint was as a "Target Study" officer at Seymour-Johnson AFB, North Carolina. Two-and-a-half years were spent in Bismarck, North Dakota, where he was the Operations Officer at a radar site. He retired from that assignment.

His son, Mark, served in the U.S. Coast Guard as a navigator on C-130s in Alaska, and also served on a boat "chasing druggies." Mark worked on electronics for a while with Boeing Aircraft Company in Pennsylvania. He currently works for GKN Company in Tallassee, Alabama. A daughter, Terri, is a Pharmaceutical Technician in Montgomery, near Gene and Betty.[1]

The Connor family members were and are proud to serve our country in any way possible, as evidenced by their longstanding tradition. One of the Air Force's core values is "Service Before Self," and the father-son Connor combo exemplifies that value. They serve(d) their country well and are to be commended.

L-R Radar Nav—G. Connor; Gunner—McClendon; Co-Pilot—Castleman; EWO D. Tihomirov; Nav—Wheeler; Aircraft Commander—Dick Charles

ROBERT MCCARTY

"When the VC ran out, we just killed 'em like killing snakes."

Private Robert McCarty aboard modified M-113 flame thrower

The Army In Which I Should Like To Fight:

"I'd like to have two Armies—one for display, with lovely guns, tanks, little Soldiers, staffs, distinguished and doddering Generals and real little regimental officers, who would be deeply concerned over their General's bowel movements or their Colonel's piles; an Army that would be shown for a modest fee on every fairground in the country.

"The other would be the REAL ONE, composed entirely of young enthusiasts in camouflage uniforms, who would not be put on display but from whom impossible efforts would be demanded and to whom all sorts of tricks would be taught. That's the Army in which I should like to fight."

Jean Larteguy (French Commando/Soldier/Journalist)

Staff Sergeant Robert McCarty was born in Opp, Alabama in 1946, the son of Charlie and Trudy (Gilmore) McCarty. If he had been in one of Jean Larteguy's armies, it definitely would have been the latter.

He grew up in the Onycha Community a few miles south of Opp and attended rural Blue Springs School through the 8th grade. His schooling stopped there, but he eventually would get a high school diploma and complete two years of college.

In 1963, McCarty joined the local Army National Guard unit, completed his AIT Training at Fort Jackson, South Carolina, then went straight to active duty in 1964 with the U.S. Army. After Fort Jackson, his Armor training took place at Fort Knox, Kentucky. He would stay with the Army for 20 years active duty, and work another 15 as a civil servant.

He was in A Troop, 1st Engineer Brigade, attached to the 1/4 Cavalry, which was assigned to the 1st Infantry Division while he served in Vietnam. His Military Operational Specialty was 11E—Tank Operator, until the last two years of his military career, when it was 76P—Supply Specialist. He also later taught Land Navigation on a tour at Fort Knox.

McCarty's first pay was $139.00 a month, $65 of which was for being in combat. His first exposure to the hazards of combat was aboard a C-130 *Hercules* flying him into South Vietnam in December, 1965, when the pilot announced they were likely to be shot at on their approach to the airfield. It dawned on him then just how serious the war in Vietnam had become and that he might be in for a rough ride.

He and three other Army Privates made up a 4-man crew who operated a tracked flame thrower which was a modification of an M-113 personnel carrier. Flame thrower duty was by volunteers only. They had a gunner, as well as two aft gunners for rear protection. The gunner operated a total of four 60-mm machine guns (two guns on each side that were actually joined together so that all four could be swung side-to-side as one entity).

He was on top of 200 gallons of Napalm under 300 pounds of pressure. As a precaution, they packed 30-40 pounds of C-4 explosives with the vehicle. That way, in case of a hit, the explosion from the C-4 would be so horrendous that the track operator wouldn't know what hit him—no long-term suffering from burning Napalm for the operator.

McCarty was part of several operations, including around Cead and north of Saigon. It was during *Operation Birmingham,* a brigade-size maneuver, that a taped conversation with Robert was broadcast over WAMI Radio Station in our hometown in 1966. I heard the airing and remember it vividly.

Discussing the ongoing operation when describing a serious ambush and firefight, he stated that, "When the Viet Cong ran out, we just killed 'em like killing snakes." That may sound harsh, but that's what happens in war—it's a matter of self-preservation, and McCarty and his fellow soldiers did what they were trained to do, and had to do, to survive.

One of the things that always haunts him to this day was having to kill young children around 10-11 years of age when they attacked his vehicle with hand grenades. The Viet Cong did not hesitate to arm and use preteen children, some of whom had waist belts with grenades hanging on the belts. He stated they sometimes tried to climb on his moving vehicle and drop a grenade down the hatch.

He recalled during the *Birmingham* operation, they started at *Phu Loi* fire-camp and moved to join up with a sister unit at *Lai Khe*. A dire situation developed when the tank convoy left its compound on a route that had plantation rubber trees on each side. Going out the gate, there were numerous Vietnamese soldiers with weapons alongside the road, smiling and waving. The American officers thought they were "friendlies."

As soon as the last vehicle left the gate, all hell broke loose as the thought-to-be "friendlies" began pouring out of the tree line and attacking with weapons such as the 57-mm Recoilless Rifle which fired a shell like a bazooka and could disable a tank with an explosive round about the size of a baseball (shrapnel flying around inside the tank was actually what killed the occupants).

The rear vehicle was hit and disabled, then the front vehicle disabled. The entire convoy was trapped. McCarty and his crew—and the others in the convoy—had to fight furiously for more than three hours just to survive.

McCarty and his flame throwers swung into action and threw a length of flame some 200 meters out, swinging in a circle and cutting a burning swath through the enemy troops and the rubber trees. Many enemy were incinerated.

The convoy commander called for Air Force CAS (close-air-support), and the jets rolled in with machine-gun fire and dropped numerous 500-pound bombs on the enemy in the ditches next to the road and along the tree line. He is convinced that without those skilled Air Force combat pilots, he would not be alive today.

The driver operated the flame thrower, and there was a gunner, as well as two rear gunners to protect aft. One of the gunners on his vehicles was shot in the back of the head and killed that day. Robert was hit in the nape of the neck with shrapnel that a medic later removed with the soothing words, "Don't worry, you'll be o.k."

His main job was to protect the commander's tank which had all the communications equipment—the CO's vehicle had all the antennae sticking up. During an ambush, the enemy would swarm that tank to try to cripple the unit's ability to maintain unit cohesion or stop the CO from calling in reinforcements. McCarty's task was to "clean off" any enemy that were swarming over the CO's tank, and, with his flame thrower, he did his job well.

McCarty never got the Purple Heart from the shrapnel. He said the fighting was so intense and so frequent that no one had time to stop and write up paperwork for accolades. They were in survival mode, and getting an award was the last thing on their minds. Their rewards were their lives. He wasn't sure how many were killed, and how many were injured, but the total was staggering.

His outfit had one young soldier who was a *conscientious objector* and "would not pull a trigger." However, he was along with the unit and would drive a vehicle and function as the unit photographer for Army records. He was willing to do his duty to his country, but refused to kill anyone by his own hand. He often took many chances with his own life to get a good picture of the action.

McCarty also had a Vietnamese interpreter/interrogator that was a tough hombre.

When the interrogator once pulled a suspect out of an ox-drawn cart and began asking questions, he tugged on the suspect's hand, and his middle finger popped off. The contraption in the shape of a finger contained a disguised Viet Cong radio transmitter. Needless to say, the interrogator was not kind to the VC wearing the transmitter.

Robert recalled the troops in his unit had been warned not to go to the nearby village to visit the prostitutes. One fellow soldier ignored the warning, and the Americans found his body in the road the next morning, minus his .45 caliber pistol, but with spent cartridge casings scattered all around from where he had tried valiantly to defend himself.

McCarty's commander ordered him to drive his vehicle near the body, and, just to be extra cautious, they hooked a winch cable to his belt loops to remove the body. As soon as Robert engaged the winch and began to drag the body, it blew sky-high from being booby-trapped.

Guts and blood and part of the torso fell through the hatch and down on top of Private McCarty, something to this day he recalls with a vivid horror. He said he threw up quite often the next few weeks, virtually every time that image popped in his mind. He will never forget the event.

He had a best buddy named Diaz who was a Panamanian and part of a crew. When Diaz jumped off the fender of his vehicle one day, he landed squarely on top of a buried land mine. The explosion was so horrific, there was little left of his friend. That scene sticks with McCarty.

It was really tough on him because Diaz had saved his life on several occasions with his quick reactions. Robert can attest to anyone he talks with that, "War truly is a living hell." He's convinced that anyone who has never experienced it will never be able to comprehend its brutality.

Often, Vietnamese kids would run alongside his vehicle and beg for food. McCarty and his fellow flame throwers had the nickname "Zippos" which was derived from the fact they used their *Zippo* cigarette lighters to ignite the balky automatic igniters on the flamethrowers. The kids would yell "Hey, Zippo, gimme candy," which they did, when they had some available.

He talked about how tough the situation was when one had to get up every day and face the stress of another day of combat. In the Vietnam War, many American troops turned to drugs which were easily obtainable as a way to cope with those stresses.

In December, 1966, on his return stateside through California, McCarty said he was disappointed how he was ignored by most people.

Some of the more vile protestors would walk nearby, light off a fire-cracker for spite, and watch the American soldiers instinctively hit the ground as they so often had to do in combat. They were conditioned to drop at the first sound of enemy gunfire. The laughter and taunting of protesters are memories that stick with him.

During his service in the Army, McCarty married first wife, Barbara Metcalf, of Opp who passed away in 2005. Their children are Dennis and Kenneth.

Other than Vietnam, he had assignments at Ft. Benning, Ft. Riley, Ft. Ord, and Ft. Knox, plus two assignments in Germany.

Robert sometimes was a rowdy guy who loved to brawl a little. He told me he lost a stripe on numerous occasions, but, "It didn't matter much because I got it right back." He was just that good a soldier. During our interview, I mentioned that an officer friend of mine had once received an Article 15 as non-judicial punishment under the UCMJ. Robert said almost immediately, "I had one of those, too."

At the 20-year point, he was offered promotion to E-7 (Sergeant First Class), with the stipulation that he go to Germany a third time. He declined the promotion and the assignment and retired from active duty as Staff Sergeant. He then went to work in Civil Service as a heavy equipment operator at Fort Rucker, and worked there for some 15 years before retiring for good.

McCarty later married Jeffie Harrell (who had lost her husband "Buddy" Floyd). He and Jeffie live not far from Dozier. There, they enjoyed restoring a home originally built in 1902. For several years, they also enjoyed fishing at a place they had near Mobile, Alabama.

Robert has fine property with wooded areas that allows him to continue with a passionate past time he had as a young man—deer hunting and dove hunting.[1]

If a *Jean Larteguy* of today's military wanted someone who is an *enthusiast in camouflage uniforms, who would not be put on display but from whom impossible efforts would be demanded and to whom all sorts of tricks would be taught,* then I would highly recommend the Robert McCarty's of the world to fill that bill and fight in that type Army.

Robert McCarty was a good soldier who answered his country's call, and he is to be commended for his service to our nation in its time of need.

DAVID J. MYERS

"He was assigned to work as an engineering aide on the X-1 and X-15 test programs"

David Myers in uniform (Air Force photo, 1972)

Retired U.S. Air Force Lieutenant Colonel David Myers is a native of Lockeford, California, born there in 1937. He eventually attended the University of California, Berkeley, where he studied Engineering.

As a youngster, he was always fascinated with airplanes. He built and flew dozens of balsa wood and tissue paper models. Eager to get airborne, he managed to get his first flight aboard a crop-spraying helicopter on his father's farm. That was the beginning flight of a long and illustrious career where he flew a variety of aircraft, from the simplest to the fastest and most complicated.

At Cal-Berkeley, he worked two six-month periods at the Flight Test Station, Edwards AFB, for the National Advisory Committee for Aeronautics (NACA) which was a predecessor of NASA. He was assigned to work as an engineering aide on the X-1 and X-15 test programs where he worked with some phenomenal people.

One of those was a young test pilot named Neil Armstrong, a later astronaut and the first human to walk on the moon. Others he worked with moved on to the Johnson Space Center in Houston where they developed technology used for the Apollo moon-landing program.

There was a draft in David's time in college, so he beat them to the punch by applying for and being accepted into the Air Force Aviation Cadet Program. His class of cadets was the last class ever before the program was disbanded. Those who know their military history are aware that the cadet program provided almost all the Army Air Corps and Air Force pilots from World War I through the Korean War.

A hot-shot student pilot, Myers graduated second in his class and was given an assignment to fly the Convair F-102 *Delta Dagger* fighter interceptor with Air Defense Command. The F-102 was capable of climbing at more than 17,000 feet per minute and velocities beyond the speed of sound. The crews could be scrambled to be airborne within five minutes of being alerted to meet any incoming enemy threat.

Stationed at Truax Field, Madison, Wisconsin, he once participated in a unique exercise where all U.S. commercial and general aviation aircraft were grounded for an entire day while mock air battles raged against simulated incoming Russian bombers. Myers was on our front line of defense, and he acquitted himself well at the height of the Cold War. He again proved his mettle a year later by flying day and night combat air patrols over the Florida Straits between Key West and Cuba during the Cuban Missile Crisis in the Kennedy Administration years.

Enduring one of the coldest locations possible, Myers served a tour at Thule, Greenland during a time when B-52 bombers flew on station, overhead 24/7. One scramble was at midnight after a B-52 crew reported seeing a UFO, which disappeared by the time he ran the intercept. The same thing happened a few nights later.

Getting a reprieve from the cold, Myers was assigned to the 331st Fighter Interceptor Squadron at Webb AFB, TX, and then on to the 319th FIS at Homestead AFB, FL, south of Miami.

On one memorable deployment while with the 331st, he was sent to Jordan to train Jordanian Air Force pilots to fly F-104s which Jordan had planned to purchase from the U.S. The training was short-circuited as the 1967 *Six-Day War* with Israel kicked off.

His squadron got their aircraft and crews out of Jordan only one day ahead of the start of the war. When they went back to retrieve ground support equipment after the war, most of it was heavily damaged.

While he was assigned at Homestead, a Cuban MIG pilot defected. Myers and the other F-104 pilots took a look at the Russian-built MIG and were surprised at the sometimes crude, but effective, design which included solid rubber tires and mechanical gauges.

At both squadrons there, he worked with aircraft maintenance, flying most of the functional check flights after daily maintenance. Mach-2 flights in the F-104 were daily fare, as it was the fastest interceptor in the Air Force inventory at the time.

Going from 10 years of flying a high performance supersonic aircraft to a small single-engine, propeller driven, aircraft that could fly about 100 mph was a huge change, but Myers successfully made the change when he was assigned as an 0-1 *Bird Dog* forward air control pilot during a 1970-71 tour in Vietnam.

He was with the 21st Tactical Air Support Squadron at Bien Hoa, and most of his time was flown zig-zagging across central South Vietnam, checking on forward operating locations where three of four pilots with their aircraft were stationed with a small Army detachment. In some cases, the only runway available was the main street of a Vietnamese village.

Enemy rocket attacks occurred often, with sirens sounding day and night. One morning after an attack, Myers went to work to find a piece of shrapnel from a rocket had punched a hole in his desk.

It was during this tour that he survived the only crash of his career, when a large Army helicopter lifted off right next to the runway during his practice touch-and-go landing and flipped his light aircraft out of control, causing it to land on its nose in a ditch. Dusting himself off, he was back in the air in no time.

He returned from Vietnam to the 14th Air Training Wing at Columbus AFB, MS, serving as a maintenance officer and performing functional check flights on T-38s and T-37s. He later became Organizational Maintenance Squadron Commander with responsibility for over 300 officer and enlisted personnel.

Myers had a tour of duty at Sikes Airport in Crestview, FL, with Fairchild Aviation, then moved down the road to Eglin AFB to head up the Quality Assurance office for aircraft maintenance in the 3246 Test Wing.

There, he flew functional check flights in the T-38, but also flew *photo chase* for test missions with a photographer in the back seat. They flew close formation with F-15 and F-16 fighters, photographing them as they dropped their bombs, sometimes in a 60-degree, supersonic dive, which requires quite the touch to hang in formation under those conditions. This was his last flying assignment.

He was subsequently assigned to Torrejon Airbase, Spain, with the 401st Fighter Wing, as Operations and Training Officer. He lived off base in Madrid, despite the frequent terrorist attacks by those of the

Basque Separatist Movement (ETA) and other groups of similar persuasion.

Most of the attacks were against Spanish military and national police, but Americans were sometimes targeted. Several attacks occurred in Myers' neighborhood, and he narrowly missed one car bomb explosion that killed 12 Spanish policemen.

After his Torrejon assignment, he returned to Florida, this time to Hurlburt Field where he managed field testing of tactical intelligence gathering and distribution systems, with the 4442nd Tactical Control Group and the USAF Tactical Air Warfare Center.

One job that came up periodically during his career was that of accident investigator. Though sometimes gruesome, Myers felt it was very rewarding spending weeks in detective work, putting together the flight history, maintenance time lines, and aircraft bits and pieces to determine the cause of an accident to prevent a recurrence.

Lt Col David Myers retired in 1989, having flown 3,600 hours, more than 1,600 of which were in fighter aircraft. He flew 187 combat sorties.[1]

He currently resides in Crestview, FL, with his wife, Hettie, this author's sister.

David Myers the military man was quite something. David Myers *the man* is quite something more. If anything needs repairing, he likely can figure out a way to do it. I've seen him rebuild engines on old tractors. He can level up an old building and make repairs, restoring old facilities to original condition to preserve their heritage. The consummate "tinkerer," he just has a way of getting things done.

A very religious man, he taught Bible classes in the base chapels at Thule, Greenland and Bien Hoa Airbase, Vietnam. He now teaches Bible classes at his Crestview church and, without fail, prepares his lessons for Wednesday night worship.

He and Hettie go to great lengths to assist others in need, especially those elderly individuals such as a lady in DeFuniak Springs who needed assistance, or the elderly gentleman in Opp who had cancer and depended on their kindness and quite regular visits.

I'm proud to salute not only David Myers the military hero, but David Myers the wonderful American who always tries to help others and improve his community and the lives of those around him.

RODNEY JOHN EVANS

"Show me a hero, and I'll write you a tragedy." F. Scott Fitzgerald

Army Sergeant Rodney Evans in uniform[2]

Florala, Alabama, derives its name because it sits astride the Florida/Alabama state line near where U.S. Highway 331 and Florida State Road 85 intersect. The town is the site of a popular recreational area built on the shores of Lake Jackson, where elegant Cypress trees spire skyward with moss hanging from virtually every limb. A long-retired Air Force T-33—#TR-938—sits as a static display by the lake.

When I was growing up, the biggest news in Florala was the annual Masonic Day Celebration, held each 24th of June. Arts and crafts, unique foods, and a boxing arena placed on the main street drew good crowds on those hot, muggy June days. My Uncle Charlie Rushing from Samson was a tough guy, and he often boxed in one of those rings. The farthest thing from the minds of locals was that one day one of their own would be a celebrated military hero.

Army Sergeant and Medal of Honor recipient Rodney John Evans was born in far away Chelsea, Massachusetts, on July 17, 1948. He eventually became an adopted son of the John C. Evans family in Florala, which ran a dry cleaning operation.

Rodney Evans was somewhat like most other guys of his day and age, dreaming about girls and cars, according to a school age friend, David Emerson Robertson, who had many memories of Rodney.

Robertson recalled that Rodney and another adopted brother, Wyman, both had to work hard in the family business. Perhaps that's where Rodney developed his sense of duty to others.

Rodney and Wyman both joined the Army under a 'buddy" system, and both served in Vietnam. Rodney married his sweetheart, Barbara Geohagan, along the way, and pulled his first full tour of duty in the Vietnam War without incident.

During his first tour in Vietnam, he wrote letters to his friend Robertson who recalled, "Many people thought they knew him well. I wish I had kept all the letters he had written to me from 'Nam. It's funny he and his brother Wyman both were the gentlest souls. He wrote how the war affected him... how killing bothered him… I remember once he started laughing uncontrollably as we talked, and I knew that was his way of dealing with things, rather than breaking down and crying."[1]

Returning stateside, Evans was assigned to Fort Rucker, Alabama. Due to financial considerations, Barbara continued to live with her family in Florala until a terrible auto accident.

On the way to Fort Rucker to take care of some business, Barbara was killed in a horrendous car crash on September 5, 1968. Evans suffered emotionally from the loss of his wife, but eventually decided to volunteer for another tour in Vietnam as a way of coping with his loss and the stresses he was undergoing.

He confided to his friend Robertson that he felt a sense of duty to all the young, green troops going to Vietnam, since he had trained many of them. His second tour started February 19, 1969.

During this second tour, he was assigned to the 1st Battalion, 12th Cavalry Regiment, 1st Cavalry Division. On July 18, 1969, one day after his 21st birthday, in Tay Ninh Province, Evans immortalized himself in military history by placing himself between a land mine and other soldiers in his unit, giving his life in the process of saving others.

On April 22, 1971, when President Richard Nixon presented the posthumous Medal of Honor to the Evans family, the following description of the actions of Sergeant Rodney Evans was read:

For conspicuous gallantry and intrepidity in action at the risk of his life above and beyond the call of duty. Sgt. Evans distinguished himself by extraordinary heroism while serving as a squad leader in a reconnaissance sweep through heavy vegetation to reconnoiter a strong enemy position. As the force approached a well-defined trail, the platoon scout warned that the trail was booby-trapped. Sgt. Evans led his squad on a route parallel to the trail. The force had started to move forward when a nearby squad was hit by the blast of a concealed mine. Looking to his right, Sgt. Evans saw a second enemy device. With complete disregard for his safety he shouted a warning to his men, dived to the ground and crawled toward the mine. Just as he reached it, an enemy soldier detonated the explosive and Sgt. Evans absorbed the full impact with his body. His gallant and selfless action saved his comrades from probable death or injury and served as an inspiration to his entire unit. Sgt. Evans' gallantry in action at the cost of his life were in keeping with the highest traditions of the military service and reflect great credit upon himself, his unit, and the U.S. Army.[3]

Sergeant Rodney Evans is the only Medal of Honor recipient in the long and storied history of Covington County, Alabama servicemen and women.

His name is on Panel 20, Row 014 on the Vietnam Wall in Washington, D.C. Evans is buried in Liberty Hill Cemetery, Florala, Alabama.[4]

As beachgoers head down U.S. Highway 331, about a mile prior to downtown Florala is a narrow street that turns east. It has been re-named Rodney J. Evans Drive, in his honor. A conference center in a state park there has also been named in his honor.

JIM AND PATRICIA POCHUREK

"Jim was one of Harry's Hog Haulers"

From left: Patricia, Author, and Lt Col (ret) Jim Pochurek (Personal Photo Oct, 2012)

During an October, 2012 gathering of *Air Commandos* at an Air Commando Association (ACA) reunion in Fort Walton Beach, Florida, I had the pleasure of becoming acquainted with Jim and Patricia Pochurek, a couple who made tremendous contributions to our country during the Vietnam War era.

Jim, a retired U.S. Air Force lieutenant colonel, was a much-decorated pilot who spent 21 years in the service and flew numerous combat aircraft, including the F-89, F-94, UC-123 and several others.

In the early years of the Vietnam War, he was a participant in the *Jungle Jim* program, *Ranch Hand* program (in 1965), and many other aspects of counterinsurgency programs in an effort to hold off the communist advance in Southeast Asia. He was with the 311th Air Commando Squadron of Colonel Harry Howton, first at Danang, then at Tan Son Nhut. The squadron was part of the 315th Air Commando Group, and, while flying cargo C-123s, Jim was jokingly referred to as one of "Harry's Hog Haulers."[1]

When flying UC-123s, the mission was to defoliate heavy vegetation to deny the Viet Cong continued safety of their traditional stronghold of Vietnam's thick forests and jungles. From January to June 1963, the squadron had the added mission to move cargo, munitions, and personnel within South Vietnam.

Serving two-and-one-half years in Vietnam, Jim flew more than 740 combat missions. He received the Purple Heart, a Silver Star, two Distinguished Flying Crosses, 32 Air Medals, a Bronze Star, Vietnamese Wings, and other military awards.[2]

Patricia (Goffette) Pochurek revealed that, while in Saigon at the same time as Jim, she had worked for famed U.S. Air Force Major General Edward Geary Lansdale, one of the foremost counterinsurgency experts ever.

Lansdale was a graduate of UCLA and one who had served in the Office of Strategic Services (OSS) during World War II, then became an intelligence officer in the Army before becoming part of the U.S. Air Force.

The OSS was the predecessor to our present-day CIA, and operated behind-the-lines programs where 3-man *Jedburgh Teams* covertly jumped in behind German lines to work with the French *Macquis* saboteurs who were so instrumental in weakening the Nazi war machine in Europe.

Assigned to Saigon's American Embassy as a Foreign Service Officer, Patricia worked closely with Lansdale, the man who after World War II had helped Philippine Secretary of National Defense Ramon Magsaysay put down a communist-inspired rebellion of the *Hukbong Bayan Laban sa Hapon*, or Huks (pronounced "hooks").

The Huks group had first been formed as a communist guerrilla group in 1942 to oppose the Japanese Army that was occupying the Philippines at the time.

In the Philippines, Lansdale and Magsaysay had been a formidable team: "Of the overall total of 25,000 Filipinos who fought in Huk ranks in the 1950-55 period, Philippine Army records show that 6,874 Huks were killed, and 4,702 were captured, while in the same period 9,458 *Huks surrendered.* The remainder stayed at large."[3]

Later in Saigon until 1963, with Patricia at his side, Lansdale was helping South Vietnam's President Ngo Dinh Diem craft a similar plan to oppose the Communist North Vietnamese. Lansdale was an opportunist who always found a way to make something of public events, as noted in his book when discussing Social-Action Committee selections in the cities:

"CANDY: Social action committee elections in the Saigon-Cholon neighborhoods acquired a festive air when Hanh and I invited Pat Byrne and Anita Lauve to watch several of them with us. The girls were foreign service officers in the political section of the American Embassy. Noticing the great number

of children who would gather to watch the proceedings, the girls brought large jars of hard candy and passed them around among the youngsters. The elders, not to be outdone in hospitality, invited us into their homes for tea." [4]

There were a lot of other inspirational folks that worked with Patricia, including Lou Conien, Colonel Val, Colonel Sam Kerrick, and Daniel Ellsberg of "The Pentagon Papers" fame.

It was in Saigon that Jim met Patricia, and they've made a wonderful team.

Now in Florida in their retired life, they travel a lot and enjoy their 45 years of marriage. Jim is a writer on travel subjects and is a member of the International Travel Writers Association, as well as a member of the Florida Outdoor Writers Association.

On one memorable trip, Jim and a group rode Harley-Davidson motorcycles on a trip from Saigon (now Ho Chi Minh City) to Hanoi. *National Geographic* covered the trip, which gained a lot of attention regarding one who had fought in the Vietnam War, then returned for a leisure trip from one end of the country to the other. He and the group on the trip were well-treated by the populace.

The Pochureks now live in Palm Harbor, just south of Tarpon Springs, Florida, and spend a lot of their time raising funds for Wounded Warrior projects, including proceeds from multi-family garage sales and matching funds that yield significant contributions to our soldiers, sailors, airmen, and Marines who come home with permanent damage from wounds that affect their ability to function.[5]

A grateful nation salutes the Pochureks for their longstanding support to our country's national defense efforts.

CHARLES "CHUCK" MANIKAS

"Aboard the C-130, he was ordered to make clean water the entire flight from Dover to Nicaragua"

Chuck with water truck at Camp Alpha in Egypt (Personal photo)

Charles "Chuck" Manikas was born in Manchester, New Hampshire on May 11, 1943, to George and Olge Manikas. At age five, the family moved to Detroit, and it was there that Chuck attended school until he decided to join the U.S. Air Force.

In October 1960, at age 17, Chuck signed up and attended basic training at Lackland AFB, Texas. Near graduation, a trainer who played the comic started handing out plastic figurines such as toy firemen hats for those who were going to become a fireman. Chuck drew a toy golden shovel which meant he "would be shoveling sh*t," in the words of the comical assignments manager.

In actuality, he became a 135-pound Water and Waste Technician whose first assignment was at Eielson AFB, Alaska, about 26 miles southeast of Fairbanks. It was there that he learned about the trade from individuals who took him under their wing and taught him most of what he learned through on-the-job training.

Ed Rolfe (a Walter Brennan look-alike), a GS-15 named Robert Miller, and a fellow named Bob Probert taught him not only about his job, but about many local things such as running dogsleds, building cabins, hunting, trapping and skinning animals, to go along with cleaning toilets.

Miller once told him, "I never ask a person to do something I won't do" as he rolled up his sleeves and showed Chuck how a GS-15 cleaned a commode. Chuck considered this the most valuable lesson he learned in the Air Force—leadership by example!

After leaving Eielson, he was sent to Minot, North Dakota, for 18 months in the Civil Engineering department, working in sewage treatment plant maintenance. While there he then got into Minuteman Missiles and went to a Boeing school so he could install water treatments for missile sites' water systems.

In 1964-65 he had a break in service to attend one year of water and waste school of the National Institute of Safety and Health (NIOSH) in Missouri. He re-enlisted in the Air Force in Kansas City in 1965 and headed for K.I. Sawyer AFB on the Upper Peninsula of Michigan. When he re-enlisted he made it clear he was re-joining for the express purpose of going to Vietnam.

Ten months later, the Air Force in October, 1966, honored his request and assigned him to Phan Rang, South Vietnam. Phan Rang had I Corps, a wing of Aussie aircraft, and U.S. F-4 fighter aircraft, and was about 90 miles south of the big airbase at Cam Ranh Bay. (Cam Ranh Air Base was a part of the large Cam Ranh Bay logistics facility built by the United States; it was the major military seaport used by the United States for the offloading of supplies, military equipment and as a major naval base).

He was part of the second contingent and set water points on binjo ditches at the edge of camp where there was a creek out back. He later set the water points about 30 miles away, and remembered that the 1st ROK (Republic of Korea) Marines guarded the place. On the back side were elements of the 101st Airborne that processed Army troops coming in country and going back stateside. Also a small field hospital.

There were occasional attempted infiltrations by Viet Cong who got through the barbed wire and tripped sensors that set off flares and alarms. The ROK were extremely efficient in dealing with threats to base security.

On one occasion the VC lobbed mortars in and killed two ROK Marines. The ROK left the base, went up on the hill, killed several VC, and then strung them up on the gate posts out front for several days as a reminder/warning to the VC about mounting any future attacks on them.

The gate led out to a main route that headed down to an R and R spot, so the hanging bodies got a lot of attention.

When the bodies began smelling terribly bad after a few days, the camp commander made the ROK take them down and dispose of the dead, but not before they got their point across to the bad guys.

It was usually safe to leave the base in daylight hours, but they were required to be back on base by 4 p.m. There were occasions when he and others "shot blindly" at dark figures near the water points that were believed to be threats, and on occasions they ran for bunkers when mortars came in.

He remembered one occasion when a Vietnamese rode in on a motorcycle to a strip where loose women hung out, got into an argument, and shot and killed two G.I.'s. There were other instances of "booby-trapped" cigarette lighters.

At a bar one had to be careful to not lay a lighter on the bar and turn your head, because there had been instances of some kid sympathetic to the V.C. grabbing a G.I.'s lighter when he wasn't looking and replacing it on the bar with a booby-trapped lighter that would blow off a finger, or worse.

Most times in 1966 he wore shorts and light shirts for duty, with no flak vests as American soldiers are required to wear these days in Afghanistan or other hot spots.

After Vietnam, his next assignment was to Marron, Spain, where he spent four years of duty. Then he was on to Zweibrucken Air Base, Germany which was a short tour of less than a year. Then in 1972 he was on to Dover AFB, Delaware for two years. It was there that he had another interesting experience.

Two days before Christmas, Nicaragua had a terrible earthquake. For 16 years Chuck had been on telephone standby to go anywhere in the world. At 11:00 p.m. he got a call and was told to be packed and ready to go and be at Base Operations by midnight.

When a C-130 arrived, he was surprised that he was the only passenger to be on-loaded. Onboard the C-130 was an *Urdulator* (water purification unit), a device in which "water is pumped through sophisticated filtering equipment resembling the cream-separating machinery found on many dairy farms. The water passes through a combination of calcium hypo-chloride, ferric chloride, and snow-white diamomacheous earth before it's considered ready for general use."[1]

Aboard the C-130, he was ordered to make clean water the entire flight from Dover to Nicaragua so that American people there could have clean water and assist with the recovery efforts of American forces (there was no bottled water those days, hence the rush job on clean water!)

On other occasions he pulled duty in Egypt at a *secret* American base in Egypt (down past Luxor in eastern Egypt) at a location referred to as *Camp Alpha* (a spot with which I am very familiar since this author was there commanding an AC-130 *Spectre* gunship in April, 1980, as part of the attempt to rescue American hostages held in Iran).

A great deal of the facility was built some 60 feet underground to avoid the daytime heat. He remembered the Egyptians had phony gun emplacements around the base, and while there, he was involved with many water projects, including one where he converted an old Russian tower to pump water from 175 miles away (from the Aswan Dam Reservoir).

The base had been on water rationing before his arrival, but Chuck solved many of the base's problems through his ingenuity and employment of techniques he had learned in his first assignment in Alaska.

Before he left Camp Alpha, he found water and sent USAFE a request that they come and drill for a permanent source of underground water to alleviate the base's shortages (This was in the time before Anwar Sadat's assassination).

He became good friends with the Egyptian base commander who had a Greek wife (Chuck himself has a strong Greek lineage, so they had something in common). He endeared himself to the commander and family after he once rushed to help their young son who had scraped himself quite badly when he entangled his flowing garment in a bicycle chain.

He and others also placed a big cross on a little chapel spot they used. This created quite a controversy when it was discovered through a satellite overhead photograph that the cross was displayed prominently in a heavily Muslim country.

He also built a swimming pool while there, and received a Meritorious Service Medal upon leaving the base.

The last ten years of Chuck's career were spent at Tyndall AFB in Panama City, Florida, until his retirement in 1986. He did a lot of research and development on hazardous waste, industrial waste, water quality, and many other things related to environmental protection.

He developed Polychlorinated Biphenyl (PCB) testing kits for sampling soil in the area surrounding power poles. Big, older electrical transformers often contain PCBs, so when they occasionally explode, there is a danger of soil contamination, and a need existed to be able to test that soil, and, if necessary, dig it out and replace it.

Chuck became an expert on Mercury spills, Hydrazine, Cyanide, Dioxins, and almost any other chemical that if spilled posed a great danger to the groundwater.

Chuck married an English gal, Elaine Goodall, of Sheffield. They have one daughter who lives in Montgomery with their one grandson.

While in Panama City, Chuck owned "The Water Trough" club and was also a club manager for the Navy Coastal System (the "Long Glass" Club for their MWR programs). Elaine ran a flower shop there for some 15 years.

Now in Montgomery, Chuck Manikas is active with Disabled American Veterans as well as the Vietnam Veterans Association (VVA). He was the VVA Veteran of the Year for 2012 because of his long-standing support and efforts to improve the lot of Vietnam Veterans.

Chuck Manikas retired from the Air Force as a master sergeant, but his contributions to our nation over a 28-year-span were befitting one of much greater recognition.[2]

He indeed is another in this group of stories about ordinary citizens who performed extraordinary service for their country in its time of need.

We never tire of honoring our war-fighting heroes, and we should always do so, but let's not forget those thousands of men and women such as Chuck who toil in the background and perform those support services without which our front-line fighters could not function as a deterrent force. We salute Chuck and his family for a job well done!

JAMES AND BARBARA JARRETT

"In combat, he took a lot of hits, but was never shot down"

Captain Jim Jarrett

James "Jim" Jarrett, Jr. was born October 24, 1937, the son of James and Bessie Irene (Tomlinson) Jarrett of Maplesville, Alabama. His father had two tours in the Army, the first in the 1930s in the Army Air Corps as a Radio Operator on *Keystone* bombers.

His father left the Army, but couldn't find a job, then stood in long lines to get back in the Army for a second tour. Most of his father's later life was spent with Bell Telephone Co., but involved a lot of moving from Montgomery to other places such as Marion, Brewton, and Auburn, Alabama, before moving to Cocoa and Zephyr Hills, Florida, where he had his own company. Per Jim, Zephyr Hills only had seven phones in the entire town when they first moved there.

As a child in Montgomery, Jim Jr. remembered that right after the Pearl Harbor attack, air raid wardens patrolled homes in his Riverside Heights housing area near Maxwell Field and scolded any family that had any shred of light showing through the windows at night. The wardens wore helmet liners and armbands, and carried flashlights.

He also recalled himself as a young lad marching alongside airmen from Maxwell who marched in formation by their housing area. He held a stick high on his shoulder and pretended to be marching in military formation with them. He idolized the men in uniform.

Jarrett was always interested in aviation as a youngster. He first soloed in an *Aircoupe* aircraft when he was only 14 years old (in those days, one could solo even though they did not have an auto driver's license). He also flew aircraft such as an L-4 *Aeronca* and a *Stearman* of WWII vintage.

Jarrett was drafted as he graduated high school in 1956 and was trained at Fort Monmouth, N.J. as an Army Signal Corps Radio Operator. His first duty was in Germany as a fixed station operator where not a lot was happening. He often changed the engine oil and tires of vehicles, but not much else, while assigned to the 17th Signal Battalion at Karlsruhe (south of Manheim).

So, he exited the Army and went to work with General Telephone for five years. During this time, he was also going through divorce proceedings from his first wife.

In 1963, he came back in the Army and was sent back to Radio School again at Fort Jackson, S.C. He then pulled a 13-month tour of duty with the 7th Infantry Division in South Korea. Afterwards, he was back stateside at Fort Knox, assigned to Communications of the Armored School. Meanwhile he worked on a degree in Aeronautical Engineering with Emery-Riddle and made the rank of staff sergeant while there.

In 1965 he was sent to Vietnam for his first tour where he was assigned with the 1st Infantry Division. Returning Stateside in 1966, his board had selected him for OCS, which he started at Fort Knox in July, 1967. He was classified as being in the Armored Field, although he had never been in a tank!

On February 1, 1968, he was commissioned as a second lieutenant. His was the next-to-last OCS class to graduate from Ft. Knox, as the Army later moved the school elsewhere. He had applied for flight school and was accepted, a happy day for him.

Helicopter training was at Fort Walters, Texas, in the H-23's (called "Hiller's Tillers" by its pilots), then he was off to Fort Rucker, Alabama for training in the UH-1 A/B models. After flight school, he was off again to the Vietnam War.

Upon arrival, his 7th Squadron CO, Colonel George Crook, promptly advised him he would be the squadron S-3, not something he sought, but Crook spotted Jim's shoulder patch and knew he was a mature individual who would do a good job. The 7th Squadron was part of the 1st Air Cavalry Regiment.

At the end of this 6-month tour, he was asked what he wanted next, so he listed the following three pie-in-the-sky requests: (1) Cobra Helicopter School, (2) R&R in Hawaii, and (3) Navy Survival School in the Philippines. This drew a comment to the effect "Damn Jim, this ain't Christmas," but unexpectedly, he got all three wishes. His Cobra school was at Vung Tau for about three weeks duration.

Jim remembered getting one "pink slip" on one ride when he fired rockets high at a swamp target, then quickly nosed the Cobra over and fired two additional rockets. He was written up severely for lack of "ammo conservation" but, to no avail, gave the excuse, "That's how we always did it in my other outfits."

Jarrett experienced a harrowing event on one particular sortie. His group commander had a lot of radio frequencies he monitored and liked to keep tabs on all his helicopter pilots and what they were doing on missions.

When Jim fired on a target during a "bunker-busting mission," the commander was flying his helicopter perpendicular to and between Jim and his target. The missile went straight between the tail rotor and main rotor of his commander's helicopter, but thankfully, slipped through without exploding. Jim never knew the commander was there until he appeared in his sight line, just as he had fired the missile.

Jarrett was somewhat of a "flying hours hog" who liked to fly a lot. His goal was to accumulate 1,000 hours on that tour, but he only managed to get 953 hours combat time in the Cobra. Being scheduler didn't hurt, as he was able to often place himself on the flying schedule.

He was grounded several times by the flight surgeon because he was "getting too much time." He also flew Hueys and OH-6s when he could scrounge time in one of those helos.

In combat, he took a lot of hits, but was never shot down. He had one memorable sortie though, with a young W-1 co-pilot named John. They took a hit in the skid, and the jar from the round knocked off a "patch" from where a previous hit had occurred. He queried John several times as to whether he was o.k., but never got an answer.

Thinking John might have been hit, he came in for a landing at Chi Lang while calling for Med-Evac. When he sat the bird down, John finally squeaked out some sort of noise, then exited and ran from the aircraft.

After Jim chased him down and asked him why he never answered, he replied, "I was too scared to say anything." The Med-Evac guys were rolling in laughter on the tarmac to see such a spectacle.

This co-pilot was a W-1 who always referred to Jim as "The Sir," since he was a first lieutenant at the time, something that drove Jim crazy coming from John's mouth.

Jarrett came out of 'Nam in 1970, and that was the end of his combat. He returned to Fort Knox flying Hueys, then was off to Fort Bliss to fly Cobras, Hueys, and OH-58s. He later went to Germany in 1977 where he flew Cobras and OH-58s and was his squadron's XO. The Cobras there were the "S" models (pretty much the same as the other Cobras, but with different weapons systems).

The last year there he and three others were up for promotion to major; all four were promoted, so they had five majors in an outfit with only one major's slot. So, he headed down to Hanau (Germany, northeast of Frankfurt) to be the Airfield Commander where he flew only OH-58s for a year.

In 1980, he returned to Fort Knox in Air Cavalry in the Attack Helicopter Section of the Armored School. There were four pilots, but Jim was the only one rated Pilot in Command (PIC). Their local regulations required any sortie must have a PIC on board.

His boss wanted aviators in the uniform of the day, even if they were flying. Jim thought that was too much trouble to keep changing uniforms all day when he was flying, so he just showed up in his flight suit on fly days, which was a source of tension between Jim and his boss.

Jim retired from Fort Knox in 1982.

Along the way with his adventure, Jarrett, in 1973, met a nurse at Fort Knox that caught his fancy. He married Barbara Delaune (from southern Louisiana). Barbara felt the need to serve, so, even after she left the Army, she then joined the U.S. Air Force and was assigned to Wright-Patterson AFB in the 1991-92 timeframe. She spent three years there, then was alerted for a tour to South Korea. Before leaving, she was allowed to pick a follow-on assignment, and she chose Maxwell AFB, Alabama.

Barbara spent three years at Maxwell (which was nice as they had a son at Auburn University), then three years at MacDill AFB near Tampa. That tour was followed by four years at Dobbins AFB, GA, where she was "just about the only active duty person on base."

Barbara retired from the Air Force at the end of her tour at Dobbins, but still works at Montgomery's Veteran's Administration Hospital. The Jarretts currently reside in Pike Road, Alabama. Their son, Jeffrey, is in marketing in Charlotte, N.C.

In talking with Jim, he disclosed that he has an Aunt Mae in her '80s who was from my hometown, but now lives in Montgomery. His Aunt Mae married Jim's uncle Millard Jarrett who was a guitarist for Hank Williams, Sr.'s band. Millard never took a formal music lesson in his life. He was self-taught, but was a good enough musician to play with maybe the greatest country music composer/singer ever.[1]

Jim and Barbara Jarrett have lived a life of adventure and service, and our country is grateful to both for their contributions to our national defense over the years.

DONALD JACK NEESE

"He was awarded the Silver Star plus the Purple Heart"

Captain Donald Neese in uniform

As a teenage youngster who was taken with sports, I often looked through an older sibling's high school yearbook to admire the photos and stories of the good athletes. One of those from the Opp High School, Alabama, Class of 1958, was Donald Jack Neese, a track star and all-conference center and linebacker on the football team.

Football players were *really* tough in those days. They played both ways, and usually for the entire game. He was also a track and field champion in the shot-put, a Beta Club member, and a National Honor Society scholar.

That class had a great coach, James Lawrence Nolen, who later became Principal, Superintendant of Schools, and eventually Mayor of the town. Other quality players such as Kenny Griffin, Waylon McKinney, Joe Grimes, Bobby Moulton, Johnny Smith, Donnie Schofield, and Wiley McLeod assured they had a really fine athletic program and winning records.

Neese epitomized the scholar-athlete, and that would carry over into a highly-decorated Air Force aviation career and post-Air Force business success. If there was ever one who pulled himself up by his bootstraps and became a big success in life, it was Donald Neese.

He was born in 1940 in the little community of Babbie (between Opp and Andalusia) to Argle and Bonnie (Whitman) Neese. He was the third and last child behind an older brother and sister. His father had been a share cropper for years, but saw the futility in that line of work and eventually moved to town to work in the local cotton mills.

Neese as a youngster had worked in cotton and peanut fields and worked a stint of part-time hours at the local cotton mill. Picking cotton, stacking peanuts, and stripping cane were hard jobs he performed before he took on the mill job. This author, whose father also was a share cropper for 15 years, did some of the same type farm work as a youngster and can empathize with that plight.

As a youngster, Neese attended grade school at Westview School on the west side of town before becoming the football and track star at Opp High School.

After graduation, thinking he might have a future as a college football player, he visited the University of Alabama in 1958, the first year of Coach Paul "Bear" Bryant's tenure, and even considered Florida State Teachers College—now Florida State University. He decided to bypass college and college football, for the time being.

After high school, Neese worked for Sonbeam Furniture Company on Holt Street in Montgomery, occasionally acting as a "repo" man when customers failed to pay for purchases. He was aggressive and was even arrested once for a repo effort when the lady wouldn't make payments on a gas range. The company owner hired a lawyer and bailed him out.

At one point, he applied for the Aviation Cadet Program, but after a call and a job offer, he had gone back to Opp and worked for Candy Johnson of Danley Furniture Co. for about six months, despite having been accepted for aviation training. On January 1, 1959, he married his high school sweetheart, Ann Crosby, in a small ceremony.

Tiring of the furniture job, in July that year, Neese on a lark joined the Air Force as an enlisted man and was told that after graduation he would be trained in Intelligence. He later learned he would fly in the rear end of a C-130 aircraft collecting and deciphering signals and audio intelligence against some of America's adversaries.

After Basic Training, he was sent to Yale University—heady stuff for a young man from South Alabama—and there learned to speak Chinese, which, unbeknownst to him at the time, proved to be a blessing that provided a stepping stone to bigger and better things. His classes were considered "immersion training," as only Chinese was spoken. There he earned 32 college credits, which came in handy later.

After Yale, he and Ann headed to San Angelo, Texas, to Goodfellow AFB, where he was trained in the Intelligence Field, another eye opener.

In February 1960, in cold weather that he could hardly imagine, he went through Survival School at Stead, Nevada, and thought the training was about as realistic as actually being in a real survival situation. The training was in and around the Donner Pass near Reno and Truckee, California. With a subsequent overseas, unaccompanied assignment, he flew out of Travis AFB, CA, on April 16, 1960, on a 31-hour journey with stops at Wake Island and Guam, before arriving at Tachikawa, Japan, near Yokota Air Base, his ultimate destination. There he became part of Detachment 1, 6988 Radio Squadron Mobile.

He flew and worked in the rear of the aircraft on what was termed *Comfy Echo* missions on some 100 missions averaging around 10 hours per sortie, listening and recording communication emissions from Communist China that were ultimately forwarded to the National Security Agency in Maryland. During a portion of this duty, he alternated between the bases at Kadena, Okinawa, and Misawa, Japan. He managed to get Ann over, despite the difficulties, and he did his critical collections job, while taking some classes at the University of Maryland campus, accumulating 60 semester hours toward a college degree.

While in Okinawa for a short while, he and Ann lived on a bluff overlooking the South China Sea. He often walked down to the WWII beachhead areas for snorkeling or swimming and thought about the Marines who came ashore in that area to rout the Japanese.

At the end of the tour, he was assigned to become a Drill Instructor at Basic Training at Lackland AFB in San Antonio, a job that he really did not care for, although they enjoyed the area.

Eventually, with a degree in hand, he applied for Officer Training School and made it through OTS in 1966, earning a commission as a second lieutenant. That was followed by a year of pilot training at Laredo AFB, Texas, to fulfill a lifelong dream.

Overcoming the pitfalls of wake turbulence in the flare, leaders of 4-ship formations who on night sorties turned *into* him when in echelon formation, and getting rather excited on his initial T-38 solo flight, Neese did well and finished 9th among 50 graduates in Class 67-H.

In June, 1967, only one fighter aircraft assignment was available to graduates, so he opted to become a pilot on the C-141 *Starlifter* cargo/passenger aircraft. After training at Tinker AFB in Oklahoma City, his first operational assignment was with the 41st Military Airlift Squadron, which was part of the 437 Military Airlift Wing. He spent over three years in that job at Charleston AFB, SC..

In that capacity, he flew worldwide missions to places such as Sicily, Kenya, Germany, Spain, South Africa, Ascension Islands, the Azores, et al, while occasionally ferrying high-level VIPs. Other missions were flown through Alaska to Yokota, Japan and other locations in the Orient, to include Tan Son Nhut (Saigon), Cam Rahn Bay, and Danang, Vietnam.

Flying the C-141, he braved the occasional false engine fire warning light or a rare hail shower near thunderstorms. However, after about three years, he began to get restless and wanted to test his mettle in combat, so he volunteered for re-assignment to the Pacific in the C-130 *Hercules* aircraft. After training at Little Rock and arriving at his next assignment, he described himself as "gung-ho, with a sense of mission." That attitude carried him to greater heights.

Checking out in the Pacific Theater, first at CCK in Taiwan and later at Clark Airbase, Philippines, Neese would get his chance to test his mettle before he realized it at a place called Khe Sanh, South Vietnam, where elements of the III Marine Amphibious Force, the Army's 1st Cav, some personnel of Seventh Air Force and some South Vietnamese ARVN were under siege by what was believed to be division-sized elements of the North Vietnam Army.

During the Khe Sanh siege, he and crew were forward deployed to Danang, South Vietnam. During this several weeks operation they flew 186 missions, 90 of which were in support of Lam Son 719 (into Khe Sanh). Some missions were flown in "0-0" conditions because of the fog and clouds surrounding Khe Sanh. Those flights were extremely risky.

On 8 Mar 71, piloting a C-130B, then-Captain Neese was assigned a combat-essential mission to airlift vitally needed helicopter fuel to Khe Sanh outpost. The North Vietnamese had it surrounded and had lobbed artillery, rockets and mortars onto the site for several consecutive weeks. Some rounds damaged the runway surface.

After landing at the short assault strip, it was discovered that two tires on his aircraft had been cut due to taxiing on the jagged ramp. With complete disregard for his personal safety, Captain Neese remained with his aircraft through a heavy rocket and mortar barrage in order to get his aircraft out of the besieged airfield with minimum exposure to hostile fire. For his heroic actions, he was awarded the Distinguished Flying Cross.[1]

What was not noted in the Citation was that Neese and crew got out and changed the tire on the C-130 while under enemy fire. It's difficult enough for an aircrew to change a C-130 tire, but to do it while under fire is quite a feat!

Captain Neese displayed further heroism a short time later when, on 12 Jun 71, while flying with the 773rd Tactical Air Support Squadron at a classified location in Southeast Asia, Neese flew an emergency re-supply CDS—container delivery system—airdrop mission to besieged friendly forces which were completely surrounded by an estimated division-size enemy force.

On his approach to the drop zone, the aircraft was repeatedly struck by ground fire, one round exploding directly in front of Captain Neese and inflicting minor injury to his leg. Despite his hazardous emergency situation, Captain Neese perfectly performed an emergency recovery of his severely damaged aircraft, saving his valuable aircraft and the lives of his crew members.

For these heroics, he was awarded the Silver Star, plus he received a Purple Heart for his wound.[2]

Note: A Silver Star is the Air Force's third highest decoration for valor, behind the Air Force Cross and Medal of Honor.

Once again, the Citation didn't tell the whole story. While being shot up, his aircraft experienced a "hang-up" of the ammo and medical supplies being dropped, and, during evasive maneuvering, one engine and the wing caught fire, requiring an engine shutdown. In spite of all that, the cargo was loosened, then dropped perfectly to the friendly forces that were under siege. His was a feat of airmanship in recovering that aircraft that is rarely achieved!

Due to enemy ground fire, his C-130 aircrew had to arrive overhead the high-threat location at 14,000 feet altitude, then do a corkscrew-like spiral down to only 600 feet before rapidly aligning the aircraft for final drop of the cargo.

He also had on occasion flown "Klong" missions in Thailand, a mission with which this author is very familiar, having flown similar routes when I was there in the mid-1970s.

A new assignment brought him back Stateside in the Spring of 1971. He was assigned to the Air Staff, as a part of the Assistant Chief of Staff, Intelligence (ACSI), and worked in Human Intelligence (HUMINT) at Ft. Belvoir, Virginia, which was an Army post, but was home to the 7602d Air Intelligence Group..

There, he supported President Richard Nixon and subsequent administrations as an escort and interpreter on numerous missions with the Chinese; he associated with the likes of Henry Kissinger, Richard

Holbrooke, and many others of influence in the Nixon Administration. He recalled that Nixon was a tough, demanding personality to work around, but he always had a lot of respect for Nixon's foreign policy genius.

In 1972 while with Air Force Intelligence, he negotiated a "Shotgun" agreement with the Soviets whereby an American pilot and navigator would "ride shotgun" on their aircraft coming into our airspace, and the Soviets would reciprocate. After the *Shangai Communique*, Neese negotiated that same type agreement with the Chinese government. He completed his military career as a Case Officer with the CIA where he handled defectors, among other duties, with the CIA's Directorate of Operations from 1976-1979.

In that job, he de-briefed Phil Smith, a prisoner of war held inside Red China for seven years during the Vietnam War-era who eventually came out of China. A book, "Journey Into Darkness," was later published about Smith's time as a POW.

Neese also debriefed Soviet defector Yuri Safronov, a pilot. Safronov jumped the border and flew into Iran. The Iranians gave him up to our CIA for debriefing.

Together, they toured numerous Air Force bases to talk to pilots and students. One of those was Maxwell AFB in Montgomery. It was in Montgomery that Safronov asked Neese to teach him to drive a car, which he did. He also wanted blue jeans, so he helped him buy blue jeans at a K-Mart!

After an illustrious military career where he had accumulated over 5,000 hours flying time, with more than 600 combat missions, achieved Senior Pilot status, been awarded some of our nation's highest decorations, and contributed greatly to our country on the Air Staff and at the White House, Donald Neese retired from the Air Force and went to work for Lockheed Corporation in 1979.

Retiring from the Air Force he so loved was a hard decision, for he had a line number for promotion to lieutenant colonel and a pending assignment to Command and Staff School at Newport, Rhode Island.

So much separation and emphasis on his career had strained his and Ann's marriage to the point of a divorce, so it was easier to work his way through this transition. Also, by this time, he was in a second marriage (to Melinda Jacoby) and had a new baby and a step-son. The time was right to go to work in the civilian world.

Despite the angst, he felt another great opportunity would present itself outside the Air Force, so he put in his retirement papers, sent out some 100 resumes, took Melinda with him on a final flight in a T-39 *Sabreliner*, and never looked back.

With Lockheed, Neese held many positions, including those such as Vice President, Americas; Vice President, Asia-Pacific Region; Chairman, Lockheed-China; and, Vice President, International.

He was with Lockheed for 25 years, culminating with his corporate responsibility for Western Europe, which accounts for about one half of Lockheed's six billion dollars international sales.

Before that was an assignment in the United Kingdom where, as Vice President, he was the Lockheed-Martin National Executive with responsibility for policy, marketing, public relations, sales campaigns, and all elements of the Corporation's UK presence. Under Neese's leadership and guidance, the Corporation's sales increased by ten-fold to over one billion dollars a year.

Other benefits accrued. He became well acquainted with the British Royal Family, becoming friends with Prince Charles and Princess Diana. He attended social events with the Queen of England in attendance.

Donald Neese was a long way from the peanut fields and cotton mills of Opp, Alabama. His success had come through much dedication and hard work, and is an example of what can be achieved by those who are "not born with a 'silver spoon' in their mouth."

After his 25 years with Lockheed-Martin, he founded and became President of *The Neese Group*, which was formed in 2004 to offer aerospace and defense advisory services to a select group of clients, including Lockheed-Martin Corporation. The Group has global experience and reach, but focuses on Europe and the UK.

He is presently the Chairman of the British American Business Association, and was among the founders and is Chairman of the German American Chamber of Commerce. He is involved as well with many other organizations supporting international trade and cooperation.

Neese resides in Virginia with wife, Melinda, and lives on a farm. They have three daughters and two grandchildren.[3]

Citizens of the State of Alabama, Covington County, and our shared hometown of Opp are justifiably proud of a native son such as Donald Neese who has contributed so much to this country's national defense and well being, as well as fostering close ties and better relations with our international Allies. All Americans salute him for a job well done.

HERSCHEL SCOTT MORGAN

"Alone, unarmed and unafraid...but Morgan felt a little unease"

Colonel (Ret) Scotty Morgan (author's personal photo 13 April, 2013)

Retired Air Force Colonel Herschel Scott Morgan was born January 14, 1934, in Candler, North Carolina, to Norman Israel and Pauline Elizabeth (Sams) Morgan. He was the oldest of five siblings (three sisters and one brother).

After high school, Morgan attended Western Carolina University. He entered the Aviation Cadet Program in July, 1954, at Spence Airbase, Moultrie, GA, where he flew the PA-18 for 20 hours (in a program designed to separate out those with little aptitude for flying), as well as the T-6 *Texan*. His was the last cadet class at Spence, with his training completed in February, 1955.

From there, he was on to Bryan AFB, TX, from February through August, 1955, where his basic flight training was in the T-28A and T-33 *Shooting Star* for advanced training.

After Bryan, he headed to Del Rio for gunnery training in the T-33 which was armed with two .50 caliber guns. That was followed by training in the F-84 (straight-wing). Shortly he was headed off to Shaw AFB, SC, for RF-84F *recce* (reconnaissance) training, arriving in February and departing in July, 1956.

An overseas assignment followed at Yokota Airbase, Japan, where he flew the RF-86 during 1956-57 with the 6021st Composite Squadron—a squadron with RF-100s, RF-86s, T-33s, and RB-57A's, and Bill Kirk as squadron commander. From there, he was assigned to Kadena Airbase, Okinawa, commencing January 1, 1958, with the 15th Tactical Reconnaissance Squadron (with the knowledge that the squadron would soon be getting the new RF-101 *Voodoo*).

When this assignment was complete in 1959, he left Okinawa with 1,400 hours flying time, $400.00 cash money, an Austin-Healy auto, a Browning automatic shotgun, and a 30-30 rifle. Morgan also married a high-school classmate that year, and two sons, Scott and Steve, would follow.

The Air Force decided that all ground controllers of the new SAGE (Semi-Automatic Ground Environment) computerized ground defense system needed to be rated pilots, so Morgan found himself with a new assignment (from 1959-62) that required extensive training at Tyndall AFB, FL, as well as Richards-Gebaur AFB, MO.

After training, Morgan was assigned to the Portland Air Defense Sector at Corvallis, OR. He was one of the first four "Expert" rated controllers west of the Mississippi.

The two computers for the system were tube type and huge in size, and had less computer power than today's average PC. The two buildings with the computers occupied about two acres. According to the experts, if the air conditioner conked out, the temperatures in the buildings within five minutes would exceed 150 degrees!

Ever mindful of honing his flying skills, Morgan managed to get in some sorties at Klamath Falls, getting around 30 hours a month in the T-33 (although the unit had F-101's).

His next assignment was to South Korea, to an island spot called PaengNyong-Do (held by the United Nations), about 5-6 kilometers from the North Korean mainland, where American C-47s landed on the beach with supplies, but only at low tide!

Morgan had taken the assignment on condition that he would have a flying assignment of his choice once the P.Y.-Do assignment was complete. There, he was one of 35 Americans who were advisors to the South Korean soldiers who operated a radar warning system on the island. It was Northwest of Seoul and just below the 38th parallel.

He took the job with a good attitude and "with visions of sugar plums dancing through his head" regarding that promised choice follow-on flying assignment!

In January, 1963, after six months on P.Y.-Do, he headed down to Osan with the 314th Air Division Command Post and a re-qualification in the T-33. His next assignment was to the 45th Tactical Reconnaissance Squadron flying the F-101 at Misawa, Japan.

The 45th crews alternated with the 15th TRS crews in Southeast Asia, flying out of Don Muang Airbase in Bangkok under *Operation Able Mable* until 1962, when they were sent back to Tan Son Nhut Airbase in Saigon.

He was logging "combat support time" as opposed to "combat time" because, by definition, combat time could only be logged if shooting or dropping bombs. He had to go for a complete re-qualification in F-101s, and he landed at Tan Son Nhut in October, 1963, then was sent home for Christmas at Misawa in December. He continued to fly missions the rest of 1964 and early 1965, accumulating 93 combat missions before a fateful mission on 3 April, 1965.

Launching out of Tan Son Nhut, Morgan would be flying single seat on an F-101-C mission on April 3rd, 1965 that would have a life-changing effect.

A March, 1965 decision to interdict the North Vietnamese rail system south of the 20th parallel led immediately to the April 3, 1965, strike against the Thanh Hoa Railroad and Highway Bridge spanning the Song Ma River. The bridge was located three miles north of Thanh Hoa, capital of Annam Province. Lt Col Robinson Risner was designated overall mission coordinator for the attack. Risner assembled a force consisting of 79 aircraft—46 F105's, 21 F100's, 2 RF101's and 10 KC135 tankers.[1]

"Alone, unarmed and unafraid" is the motto of Air Force reconnaissance pilots, but Morgan felt a little unease before the mission started. He was initially supposed to be a "pathfinder" for the other fast movers, but that was changed to have him collect BDA (battle damage assessment) once the attacking aircraft completed their runs on the target. Then, his scheduled takeoff was delayed. He had a nagging feeling in the back of his mind as he picked up his helmet and parachute. When he and his wingman got to startup, the wingman's radio was inoperative, another bad omen.

This was aggravating, especially since he had a new second lieutenant for maintenance officer who stood around with his hands on his hips while a sergeant worked to fix the radio. (Morgan's secondary duty was maintenance officer, and he had a bad feeling about that new lieutenant. He figured he'd deal with it when he returned from the sortie).

When they finally taxied out, the last three items to check were "pins, canopy, lanyards (for parachute D-ring)." When he reached down and checked "pin" for the gold pin in his lap-belt, he realized he had missed closing the pin earlier. He caught it this time, and that correction would ultimately save his life that fateful day.

Other things were not so routine. He and his wingman had to fly faster to get to "Yankee" (a point overwater at the DMZ) to get fuel from the tanker, so they sped it up.

Colonel Robbie Risner relayed there was "lots of flak at 8,000 feet," so they flew a little higher. When they joined their tanker for fuel, Morgan's wingman couldn't get his fuel. Another omen. Morgan coaxed him to, "Stay on the wing anyway and plan to recover at DaNang, and lead will recover at Saigon."

After the attacking aircraft went down the chute, Morgan ran in and got the BDA on the Thanh Hoa Bridge, turned and then headed back out over the coast, then turned around and used a small island as an IP (initial point) to start a run-in on a radar site about 75 miles from the bridge.

Intelligence had indicated the North Vietnamese were constructing the radar site, and a Major Rose with Intel had requested before his departure that Morgan reconnoiter the site because the major had a briefing coming up later that day for a general who was pressing to find out about the status of the site under construction. In fact, the major had given Morgan a list of five possible targets. He would never get to the second one.

At the target, he pulled up to 200 feet altitude, hit the afterburner, and was doing 515 knots (about 590 mph) when he felt three solid thumps on the airframe.

Suddenly, the rudders began to flutter, the control stick went limp, and gauges in the cockpit went crazy. Morgan quickly realized he had no control of the aircraft. He began to rotate in what he initially thought was a roll, such as an aileron roll, and saw trees when he raised his head to look out. He knew he had to get out and thought to himself, "What a hell of a way to go!"

He reached and pulled up the ejection handles then squeezed the triggers and was ejected immediately from the aircraft, losing his helmet at the same time. He was upside down going out, or at least he thought so.

His Weber Ejection System worked, his parachute opened, and as his eyes followed his crippled, flaming aircraft, he realized the F-101-C, number 560075, was actually tumbling end over end. He followed it visually until it slammed into a hillside in a horrific explosion. The F-101 had no self-sealing fuel tanks in those days, so it made quite the spectacle in its flaming descent to impact.

Morgan recalled that many years later, several searches were made, but the actual crash site on the side of the mountain has never been found.

As he came down through the trees, without thinking, he pulled the "apple" which deployed his overwater survival dinghy which was bright in color and the absolute *last thing* he needed as he began to think about escaping and evading. During his landing through trees and bushes on the side of a hill, he twisted his ankle. Later on, he realized he had probably dislocated his shoulder.

With adrenalin flowing and his mind racing, after landing, he began to formulate his plan. He got rid of and moved away from the bright dinghy, then went through his survival kit to get his emergency locator transmitter (ELT) and hand-held radio that could transmit on Guard frequency.

He had practiced numerous times extending the antenna on his ELT and remembered it had to be pulled completely out to the end of travel for it to work. In his excited state, he pulled the antenna so hard that it came completely out of the transmitter, rendering it useless. He kept it in his flight suit pocket.

Knowing that North Vietnamese were likely in the area looking for him, he opted not to use his hand-held radio for the time being. When he tried to use the emergency beacon, it just cackled, so he kept it off.

He was near a stream and could hear water running. He moved cautiously and went upstream where he hid in a bamboo thicket the first night. He heard the Vietnamese shoot the dinghy to deflate it, so he knew they were not far away. They did a lot of other random shooting as well.

In their initial search, Morgan estimated they were at one time within 10 feet of his location without spotting him. He stayed put in the thicket the first night and was amazed at the number of glow worms he observed in the darkness.

The next morning, the searchers came again, this time shooting randomly, which scared up a big animal that bolted and ran. The Vietnamese shot at it.

Around noon time, he got on a trail and moved into a big ravine where he found another thicket to spend the second night. By this time, he was very thirsty and felt the urge to get water might be clouding his judgment and affecting his caution.

The next day he moved across a mountain into a young banana grove for cover. His thirst began to be overwhelming, and he knew he had to get water. He was not hungry the first two days, but thirst was becoming a driving force in his quest to evade and survive.

He suddenly noticed he had at least 50 leeches on his body, so he brushed them all off. His survival kit had plastic bags and Iodine tablets for water purification, so he put a tablet in a bag in anticipation of getting stream water, then waited.

Near the stream, he first saw two Vietnamese women moving upstream with "choggie" sticks over their shoulders and carrying a container on each end. An armed guard was with them which heightened his need to stay silent.

After they passed, he tried to get down to the stream, but the bank was too steep, so he moved parallel and upstream to a better location. As he moved a bit, he found a small trail down to the water's edge and checked it out. On the other side of the stream was an "old gook" on a water pipe with a sword.

Seeing that, Morgan tried to back up and walked smack-dab into two women with a guard who was carrying a shotgun. The guard screamed and began shaking so hard that Morgan feared he might inadvertently shoot him with the shotgun. There was a lot of yelling, and two other militia fired rounds in the air to alert additional militia who quickly appeared. Morgan's fate was sealed, and there was no escaping this time.

Holding him at gunpoint, they removed his flight gloves and demanded he remove a gold Masonic ring he once had made by James Jewelers in Bangkok. He curled his hand to refuse, but quickly opened up to give up the ring after a woman thrust a big knife in his face. They also took his survival switch-blade knife. He never recovered either after being released from captivity. He did, however, eventually get his flight suit back.

Another eight armed militia came and took him downstream. They wouldn't let him get a drink of water, despite his efforts to communicate his great thirst. However, after walking about 30 minutes, they stopped and offered him a cigarette. A militia opened a first-aid kit, opened up a *Mercurochrome* (topical antiseptic) container, and swabbed him to treat all his insect bites. Another made his picture.

Reaching a village, they stopped him in front of a small table that had part of his parachute canopy and some small papers that had fluttered out of the tumbling aircraft.

No one spoke English, but they got him some hot tea with sugar. Morgan emptied all the sugar in the tea and swigged it down. They then offered him some beer which he refused (alcohol is a dehydrating fluid). In the meantime, more and more onlookers congregated around to gawk.

A senior military man put guards with AK-47s on either side of him and pushed him outside. Many with bamboo torches lit yelled and chanted at him, and several thrust the flaming torches in his face.

Then they started marching again, this time through rice paddies. Everyone wanted to see him, putting their torches very near his eyes. Hours later they arrived at a bigger village where he was held in a little school house (the room had blackboards for the teacher's use). They gave him a cigarette.

Next, they put him in a rowboat that had about eight inches of water in the floor, forcing him to sit in the water while they crossed a river. After the crossing, he was put on a Russian jeep camouflaged with vines sticking out at all angles. They rode and rode and rode some more. Extremely fatigued, he dozed on and off, but didn't know how much.

That night they took him down to Vinh—he could see and read the occasional road sign indicating Vinh up ahead. They spent quite some time at night waiting to put the jeep on a ferry to cross a river (the American pilots had knocked out the bridge).

After they crossed, they got on Route 1 and moved on into Vinh. That was the first time he was blindfolded as they drove around and approached what was obviously a military compound. It was raining quite a lot, and his captors made him stand under the eaves of a building with rain dripping on him. He was very tired and confused and very stressed.

They tied him to a bamboo pole, then stretched another smaller bamboo pole across his shoulders, stretched his arms to either side, and bound them to the lateral smaller pole. They then removed his blindfold in a room with a lot of light, and the interrogation began.

He first told his captors, "Just kill me, I don't give a damn." But he did. When he entered for the interrogation, he saluted the Vietnamese who did not return the salute. This pissed off Morgan, so he crossed his arms in defiance. They immediately pounced on him and beat him around. He decided crossing his arms was counter-productive.

He was asked the name of his squadron commander. He gave a smart-aleck reply "Clawed Balls." They knocked him off the stool, removed him, kicked him, clicked what sounded like a round into an AK-47, pointed it at him, and snapped off a blank round.

Then, once back inside, his captors displayed his FLIP (Flight Information Publications) charts that had fluttered out of his F-101 aircraft at ejection. In plain English was written on the charts the name *Lt Col James A. Bryant, 45 TRS, tel. #s 3668 and 2148.*

His tormentors knew the answer before ever asking the questions of him, and all because the aircraft that was supposed to be sanitized was not. Some other pilot had left things laying around, and it caused Morgan some grief.

One asked him "Parlez-vous, Francais?" wondering if he spoke French. He ignored the question. They constantly yelled at him, "You are a war criminal" and, "Do not expect to be treated in accordance with the Geneva Conventions." He would hear those statements thousands of times during his captivity.

During interrogation, on one occasion he was blind-sided by a small Vietnamese who had only one arm. Apparently he had been wounded while fighting the French before the Americans.

With his good hand, he suddenly without warning cold-cocked Morgan with such force that he saw stars. He remembered he had never been struck with such force before.

The severe torture started in May, 1966 with such methods as the "rope trick" and the "twisted handcuffs." Cuffing one's arms high on the wrist between the wrist and elbow and tight on the bone created great pain. With the arms cuffed together behind the back, if a POW moved one arm to seek relief, it only increased the pain on the other arm. His captors were masters in cruelty. He spent some two years in what the POWs termed, the "Briar Patch," some 33 miles W-NW of Hanoi. Per Morgan, there were 12 POWs there during the winter of 1965.

His captors tried to coerce him into filling out a "biography" which was simply a political document they wanted to use to exploit the captives. He was once forced to sit on an uneven, 3-legged stool for a week as they tried to force him to pick up a pen and begin filling out the document sitting on the desk in front of him.

Sometimes the physical pain was so excruciating that he had to give them some kind of answer. He would make up a lie for an answer, then spend hours later in his cell memorizing which lie he told so he would not get caught in a lie later on. That technique became an art for the POWs.

An example was one he told saying he spent three years at San Antonio AFB coaching Little League Baseball. For made-up assignment changes, he always got to a made-up assignment location on July 4th and always rotated out on Thanksgiving. That made it easier to keep track of the lies.

His tormentors always wanted him to apologize to the Vietnamese people, denounce his country, and admit to being a war criminal. Those demands never ceased.

Scotty Morgan was the fifth POW captured in the Vietnam War. Those prior to him were Everett Alvarez, Bob Shumacher, Hayden Lockhart, and Ray Vohden. He was the first to be imprisoned at what the POWs dubbed the "Heartbreak Hotel."

He recalled the year 1969 was particularly brutal. That was when POWs Ed Atterbury and John Dramesi escaped through the roof. They traveled three miles within 12 hours, but were recaptured the next day, and Atterbury was tortured to death. They almost tortured Dramesi to death:

Dramesi was put face down on a table, and while one guard held his head, two others beat him with a four foot length of rubber taken from an old automobile tire. They also slapped him repeatedly in the face. This went on for days, in ninety-minute sessions, after which the left side of Dramesi's head was swelled up like a pumpkin.

They also put Dramesi on a bread and water diet for 30 days. At other times during the next two weeks, Dramesi's arms were bound tightly together behind him and his wrists and ankles cuffed in heavy irons. A rope was looped around a two-inch-thick bar attached to his ankle irons, taken around his shoulders and his head drawn between his knees.

He was held in this position for 24 hours without sleep. His circulation impaired, the flesh on his ankles died, and he still bears the scars. After two weeks, the Vietnamese realized he might lose his feet, so they removed the irons and treated the wounds, but replaced them.

Dramesi wore the irons continuously for 6 months, removing them only once a week when allowed to wash. After 38 days of this torture, Dramesi was near death."[2]

Not only were the escapees tortured, but the other POWs such as Morgan paid a heavy price as well.

He remembered POW Bill Robinson who was recently honored with a display at the Gunter Enlisted Heritage Hall in Montgomery, AL.

Morgan said Robinson was a young crew chief—only an Airman First Class at the time—on a Kaman HH-43B *Husky* rescue helicopter from Det 3, 38 ARS, attempting to pick up Will Farby on 20 Sep 65. Farby was a downed F-105D *Thunderchief* pilot attempting to evade capture.

Robinson's helo pilot was Tom Curtis, Duane Martin was co-pilot, and Art Black was a paramedic. Morgan said they had Farby in a sling when their helicopter was taken down and its crewmembers taken prisoner.

He recalled Robinson was often forced to chop wood for cooking rice. Robinson used the chopping to communicate with other POWs via the Tap Code. Their Vietnamese captors never caught on.

The other POWs put Robinson through Officer Training School during his captivity, and the Air Force recognized that training, making him a commissioned officer after his release in 1973.

Scotty Morgan was finally released and left Hanoi on the first freedom bird that landed in the Philippine Islands on 12 Feb 73. Jeremiah Denton was first off the aircraft, then the others in the order of their capture. Morgan followed Ray Vohden off the plane. Video shows Vohden was on crutches, with Morgan deplaning behind him.

He was put through a de-briefing process and physical examinations, and then saw his 7-year-old son Steve for the first time upon his arrival at Maxwell AFB on 15 February, 1973. Steve had been born five months after Morgan was shot down and captured.

After his release, he returned to active duty with the Air Force. He first attended Air War College at Maxwell AFB to get caught up. He stayed in Montgomery for a while longer to complete degrees with Troy State University, majoring in Social Science, with minors in Business Administration and Physical Science.

He first headed back to Randolph AFB in 1974 to get checked out in the T-38. Then he was on to Bergstrom to sign in and head out to Shaw AFB for a check out in the RF-4. He went back to Bergstrom to become Chief of Operations and Training in early 1975. On 1 Jul 75, he became Commander, 91 TRS at Bergstrom. When the colonel's list came out in December, 1975 Morgan was on the promotion list.

In January, 1977 he went to Shaw AFB to become Director of Plans for 9th Air Force. General Hartinger was commander of 9th AF at the time, and Bill Kirk was the DO (Morgan and Kirk had flown F-101's at another location before).

In 1980, Morgan was assigned to a new Joint Special Operations Command (JSOC) at Ft. Bragg, N.C. where he assumed duties as J-4 with responsibilities for all command logistics, working for Army Brigadier General Richard Scholtes.

During his time there, JSOC dealt with the Grenada invasion, the hostage taking of Army BG James Dozier in Italy, the Achille Lauro incident, and the TWA 847 hijacking in Beirut.

Morgan retired 1 Oct 85 and moved to Asheville, N.C. where he first worked two years for Chemtronics, a subsidiary of Halliburton. There, he met present wife Carol. He spent another year recruiting for the company, then decided to run for Congress to represent the North Carolina 11th Congressional District.

He came in second in the Primary to Charles Taylor, but was asked to then join and campaign for Duncan "Lauch" Faircloth who was successful in winning one term in a Senate seat representing North Carolina.

He supported and worked with Congressman Faircloth for six years until Faircloth lost to another North Carolinian named John Edwards. Edwards later became a Democrat Party candidate for President before a scandal derailed his ambitions.

Scotty Morgan was a Precinct Chairman and later a Regional Director. Before that, he had assisted Senator Jesse Helms in his campaigns.

Today, he and Carol maintain a home in Asheville, but spend a few months on the road each year in their big motorhome as they enjoy the peace and quiet of such scenic spots at the Pensacola NAS campgrounds where this author caught up with him for the above interview.

Scotty Morgan paid a huge price for the defense of our country. As a POW for 7 years, 10 months, and 9 days, he endured torture that many lesser men could not have lived through. Many in his shoes never came home.

Our nation is grateful for his service before, during, and after his sacrifice in captivity, and I am personally thankful that I came to work with and know such a hero as Colonel Herschel Scott Morgan. His story proves that America truly is *The Home of the Brave.*

POST-VIETNAM/ PRE-DESERT STORM

On Feb 15th, 1944, following a night tow operation in the China-Burma-India Theater that resulted in four British and three American troops killed, the American glider commander expressed concern that the British would not have confidence flying in the gliders again.

British General Orde C. Wingate sent the following note to the 1st Air Commando Group: "Please be assured that we will go with your boys any place, anytime, anywhere."[1]

Reflecting General Wingate's confident statement, "Anytime, Anyplace" became the motto of Air Force Special Operations Commandos after the war and has continued to this day. Stories of their missions will reflect that those forces truly must be prepared to go anywhere to do an assigned mission at any time.

After Vietnam, the country was in an anti-war mood. President Gerald Ford had presided over the last dying-day actions of the Vietnam War, which included the evacuation of refugees from Saigon and a nearly-botched freeing of the SS *Mayaguez* merchant vessel that had been grabbed by Cambodians.

In this atmosphere, Jimmy Carter was elected president over Ford in 1976, and Carter's four-year presidency saw a hollowed-out military after his postponing development of new weapons systems, as well as a great deal of cutting flying hours, steaming hours, and training hours. When a series of international crises erupted in mid-1979, the military and country at large were ill-equipped to deal with them.

Anastasio Somoza was ousted by Communist *Sandinistas* in Nicaragua in the summer of 1979. About all America could do was to be on standby in the Panama Canal Zone with AC-130s and helicopters to try to protect our personnel in Managua. I was commanding one of the AC-130s, so I had many remembrances of those days.

On November 4th, 1979, Iranian militants seized the American Embassy and 66 personnel, with the result that 52, including our Ambassador, would be held hostage for 444 days.

The next month Soviet soldiers moved into Kabul and took over Afghanistan and would be there for ten years. About all we could do in retaliation was boycott the Olympics being held in Moscow.

The age of modern terrorism was upon us, and would increasingly affect our military and civilian population, both abroad and at home. Large sums of money would have to be poured into existing and newly-formed organizations such as those capable of conducting special operations and counterterrorist actions.

The *nature of warfare changed*, and our strategy, tactics, training, and preventative measures would, out of necessity, have to change for us to survive as a nation. With these factors in play, many significant military actions would occur, with many fine soldiers, sailors, airmen and Marines contributing during challenging times. Their stories follow.

Between November 1979 and April 1980, a depleted American military, with aging and ill-equipped machinery, put together an ad-hoc team of special operations ground and aerial forces to train as best they could and attempt a rescue of the American hostages in Iran on 24-25 April, 1980.

The rescue went badly as a helicopter crashed into a C-130 in the Iranian desert, with a resultant loss of eight lives. The loss of face and embarrassment to the United States government and military was even worse. Despite that, some individuals performed heroically.

One of those was Major Logan Fitch, Army Special Forces and original member of Delta Force.

MAJOR LOGAN FITCH

An Army soldier I came to admire over the years was one who grew up in circumstances similar to my own. I had a connection with him that went back to the first few months of 1980, but never met him until 25 years later, April 24, 2005, at a Ft. Walton Beach, Florida, gathering of participants in *Operation Eagle Claw*, the failed 1980 attempt to rescue American hostages held in Iran.

Fitch had many close calls in his career, but it was on that mission that he and a squad of his men came within a whisker of being burned alive. At the gathering, he briefed veterans of the mission as to his role in leading a Delta Force element into the American Embassy compound in Tehran, and I briefed the group on my role commanding an AC-130 *Spectre* gunship that would have provided overhead air cover for Delta, had the mission gone as planned.

Years later, after reading a memoir I wrote, Logan wrote back, "Jim, I just finished your book (*Airman's Odyssey*) and enjoyed it immensely. I found some interesting parallels in our lives: My father was a tenant farmer and completed the 4th grade. My mother finished the 8th. There were seven children in my family..." (my own father in Alabama had been a tenant farmer and completed the 4th grade; my mother had finished the 7th grade. We both made careers of the military, mostly in Special Forces/Special Operations. We both were part of *Eagle Claw*, and I, too, thought the similarities were uncanny).

He continued, "I am the first and only one of my generation to attend college. I enlisted in the Army at age 17 without having graduated from high school to 'get away' and sort of bummed around for a few years. Many of my officers thought I had potential and wanted me to try for OCS, but I couldn't stay out of trouble long enough."[1]

Logan Fitch went on to tell me about his early Army career. He went to Vietnam in 1965 and made E-5 rank (Sgt). He completed that tour and left in June, 1966, to attend Infantry Officer Candidate School. After pinning on second lieutenant rank and completing jump school with the obligatory five parachute jumps, his next assignment was to Berlin as part of Special Forces, despite his lack of experience in SF.

Fitch headed back to Vietnam in 1968, and, by his third tour, became a company commander in the 173d Airborne Brigade, a storied unit that had deployed to Vietnam from Okinawa in May, 1965, and was there until 1971. The unit suffered 1,606 killed in action, and 8, 435 wounded.[2]

Following that tour, Fitch was awarded a Regular commission and assigned to the Ranger Department at Fort Benning, Georgia. Then he was off to the Advanced Course, and not long thereafter, off to college at the University of Alabama.

He was given two years to complete his degree, and it was there that he met his wife Sandra (Hodges), a native of Fayette, Alabama. In the process, he became a huge fan of the Crimson Tide football team. Fitch later got a Masters degree from USC while in Europe, again with Special Forces.

In 1979, Logan Fitch was with the Army's elite counterterrorist unit, the famed *Delta Force* that had been established by Colonel Charlie Beckwith, another Southerner from Georgia, when it was given the task to train for and execute a rescue of 52 American Embassy personnel held hostage in Tehran, Iran.

An experienced soldier and one of their very best, Major Fitch was designated to lead Blue Element—one of three Delta assault teams—on the hazardous mission. Major Jim Knight was the leader of White Element, and Major Peter Schoomaker was the leader of Red Element. Peter Schoomaker would later become the 35th Chief of Staff of the U.S. Army.

The bulk of the American hostages were held in the American Embassy, but intelligence known to Captain Wade Ishimoto, Delta's Chief of Intelligence, indicated three hostages—Charge d'affaires Bruce Laingen, Victor Tomseth (a Senior Political Adviser), and Mike Howland (a Security Aide), were held separately in the Ministry of Foreign Affairs, which complicated the rescue effort.

A Joint Task Force 79-1 was formed and charged with responsibility to train aircrews and ground personnel for the hazardous mission which had long odds of succeeding. The rescue attempt plan called for a two-night operation, with 16th Special Operations Squadron AC-130s and crews from Hurlburt Field, Florida, commanded by Lt Col John Gallagher, Captain Donn Kegel, Captain Herman "Bubber" Youngblood, and this author orbiting overhead the operations, but involved only on the second night, when the Delta Force elements planned to storm the Embassy and free the hostages.

"Night One of *Operation Eagle Claw* involved Delta Force being transported by six C-130s (some carried fuel bladders inside the cargo compartment) departing from a British airbase on Masirah Island, Oman, to a transload desert site given the codename DESERT ONE for the operation.

"That site had been clandestinely surveyed by a combat controller, Air Force Major John Carney, who installed in the sand radio-call-triggered pop-up runway lighting for the C-130s' arrival in the dead of night.

"Once there, eight RH-53D helicopters flown by U.S. Marine aviators launching off the U. S. Navy aircraft carrier *Nimitz* in the Indian Ocean would land behind the C-130s, take on fuel from the C-130s, then load 120 Delta Force operators aboard the helos and proceed to a mountainous hide site about 50 miles outside Tehran where they were to be bedded down for the daylight hours awaiting Night Two operations.

"Night Two involved the Delta Force operators riding pre-arranged rented trucks into Tehran to the American Embassy location where American hostages were being held, with our AC-130s flying across the northern part of Saudi Arabia and into Iran to provide fire cover for the airfields and American troops."[3]

The rescue effort had a lot of moving parts, and when that is the case, unless things go perfectly, a phenomenon called *Murphy's Law* kicks in. In a nutshell, anything that can go wrong will go wrong. Certainly, a lot went wrong, and it almost cost Finch his life, as well as the lives of about 45 of the Delta Force operators.

The six C-130s arrived at the Desert One trans-load site, but the helicopters arrived at the site almost a half-hour late, and with only six of the original eight making it that far. They had encountered en-route sandstorms (termed *haboobs*) that caused mechanical and pilot disorientation problems, with the formation breaking up en-route. Of the six that made it, one had a serious hydraulics problem, so only five were flyable from the site on to the mountainous hide site.

Colonel Beckwith of Delta Force had pre-determined that anything less than six operational helos at the desert site constituted a no-go, so a decision was made to abort and return all forces to their original bases.

The helicopters would need fuel for the return home, and they would have to reposition behind the C-130s so as to top off their fuel tanks.

In the repositioning of one of the RH-53D helos, it crashed on top of one of the C-130s, then exploded and burned, with the resultant loss of five Air Force Special Ops aircrew members aboard the C-130. Three enlisted Marines in the helicopter also perished in the fire.

Major Logan Fitch and his Delta Force members were in the rear of that C-130 and barely escaped with their lives as they scrambled to exit via the starboard-side rear troop door which was the only escape route not engulfed with flames from burning fuel, hydraulic fluid, liquid oxygen, cooking-off ammunition, and anything else that will burn on an aircraft.

The fire was more intense by virtue of the fact that an extra fuel bladder was in the cargo compartment of the C-130 so as to provide extra fuel for the helicopters. All the helicopters had to be abandoned in the desert, and the C-130s loaded aboard all those still alive and flew them out of the desert. Eight bodies were left in the burned out wreckage, and all helicopters were abandoned.

In the haste to get away from the burning wreckage and avoid the other aircraft catching fire, a great deal of classified information was left in the helicopters. The Iranian government and its militants exploited the situation for all it was worth politically, and the United States had a lot of egg on its face. The disaster in the desert was a tremendous blow to the prestige of the United States, as well as to the Department of Defense and individual services involved.

President Carter and Secretary of Defense Harold Brown convened a special investigating team headed by Navy Admiral James Holloway, and the investigators made numerous specific findings upon which new efforts would be made to stand up new units and commands and rectify the problems with the aborted rescue mission.

Congress in 1986 would pass specific legislation known as the *Goldwater-Nichols* legislation that would provide for specific special operations organizations and missions to have their own budget and use moneys specifically set aside for special operations.

Not having to compete with conventional forces for funding by virtue of a new *Major Force Program 11*, the organizations would have specially-selected, dedicated personnel permanently-assigned units who work and train together on a continual basis, and would be able to acquire unique, state-of-the-art equipment for future operations.

By May 1, 2012, when DEVGRU (SEAL Team 6) rode specially-designed and equipped Army Special Ops helicopters into Abbottabad,

Pakistan, to take out Osama bin Laden, those re-organization and re-prioritization efforts would pay off handsomely.

After his narrow escape during the aborted rescue mission, Logan Fitch moved from Delta Force to assist some of the new units established afterwards.

Fitch later confided to me what happened in the aftermath of his Special Operations career. He said, "I then asked for a terminal assignment to ROTC. I remember an assignments officer telling me that I had a terrific combat record, but I was *just not the kind of officer to whom our young cadets should be exposed*!"[4]

Major General James Vaught (who had been the overall JTF 79-1 commander of the attempted hostage rescue in Iran) intervened, and Fitch spent his last two years in ROTC where he finished his second masters degree (MBA).

He remembered his father attended the graduation, something that was very gratifying.

Logan Fitch now resides in Forth Worth, Texas, with his wife, Sandi (to whom he's been married more than forty years), and spends time with a grandson, teaching him to fish and enjoy the more leisurely aspects of life.

They occasionally travel to Tuscaloosa to visit her parents who reside there, or sometimes to reunions. Or, he goes on trips such as one to West Point recently at the invitation of the Superintendent, U.S. Military Academy, who invited him there to share his experiences with young cadets, despite the admonition of that dip of an assignments officer who once opined that Logan Fitch was the type officer to whom younger cadets should not be exposed.

Conspicuously absent was the assignments officer who was nowhere to be found during Logan Fitch's visit to West Point.

BERT LANGFORD

"Tomcat crews maneuvered into position to shoot down two SU-22s"

Bert Langford's ice cream truck on Sandestin Beach (author photo)

The Disney song "It's A Small World" is a catchy little tune, especially when emanating from an ice cream truck along the crowded, sugary-white beaches of Sandestin, Florida, on a warm, sunny April day during spring break for many schools.

The little ditty was enough to get the attention of our grandchildren, Josh and Carrie, who made sure we were aware of the presence of the vehicle.

Bert Langford operates the truck in the warm months in this Florida Panhandle resort area that is a long way from the shores of Tripoli, Libya. And, yet, Bert has a connection to Libya that very few would suspect when he collects money from doting grandparents and hands out dream sickles, fudge sickles, and ice cream drumsticks.

Bert Langford was born in Montgomery, Alabama. His father ran a repo business that was located at the foot of the bridge crossing the Tallapoosa River between the Chisholm area of North Montgomery and Wetumpka.

Bert lived in the Dalraida area, not far from Gunter, which is part of the Maxwell-Gunter Air Force facilities. He attended local area grammar schools and graduated from Robert E. Lee High School in Montgomery in 1980.

He then joined the U.S. Navy where he was trained as a fireman and then assigned aboard the aircraft carrier USS *Nimitz*. It was during one of their cruises in the Mediterranean that a memorable incident occurred that would stick with Bert.

In 1973, Libyan leader Muammar Qaddafi had declared the adjacent Gulf of Sidra to be a "closed bay," thinking he could restrict U.S. Air Force or U.S. Navy operations through that area. For several years, tensions ran high there, not only for Navy vessels, but also for aircraft movements, as Qadaffi tried to restrict U.S. vessel movements to an area north of Latitude 32°30' which was *not* recognized internationally.

Qadaffi, who seized power in a 1969 coup, had long been an American antagonist. Each year, Libya trained some 8,000 terrorists, providing false passports, transport on Libyan airliners, and access to safe houses across Europe. Libyan support for terrorist operations exceeded all nations except Iran. It dispensed $100 million to Palestinian terrorists eager to strike Israel.[1]

In April, 1980, when I flew an AC-130 *Spectre* gunship from Hurlburt Field, Florida to an austere Egyptian base southeast of Luxor, my crew and I hugged the boundary lines of the Flight Information Region, making sure we did not put ourselves in jeopardy of a fast-moving Libyan fighter jet that could easily outmaneuver our slow-moving aircraft.

The United States was not about to allow Qaddafi to declare that part of the Mediterranean off limits to American ships and aircraft through his so-called "line of death." The U.S. recognized Libya's boundaries to be the internationally accepted 12 nautical miles offshore, but no more.

During the Reagan administration that came into office on January 20, 1981, the U.S. Navy operated in all areas of the Gulf of Sidra outside the 12-mile limit to assert safe passage through the area and test Qaddafi's resolve.

On 18 Aug 81, Qaddafi's Air Force sent up dozens of aircraft that made pass after pass against U.S. Navy aircraft and ships with fleet elements involving the USS *Forrestal* and the USS *Nimitz* aircraft carriers. That turned out to be a mistake of the first magnitude.

On 19 August, a day that Langford remembers well, two U.S. Navy Tomcat fighter aircraft shot down two Libyan Air Force *Sukhoi* SU-22 fighter aircraft. The leader of the flight of two SU-22s had flown directly to meet the fleet and had fired an AA-2 *Atoll* air-to-air missile that, fortunately, missed the Navy Tomcats. Intercepted radio calls indicated the

Libyan flight lead told his ground control that he had fired a missile at the American aircraft.

After evasive maneuvers and the near miss, the two *Tomcat* crews maneuvered into position to shoot down the two SU-22s with AIM-9L *Sidewinder* air-to-air missiles. Both Libyan pilots ejected from their crippled fighters.

One American Navy pilot said he saw one Libyan eject without his parachute opening. Other communications during the incident seemed to indicate both Libyans ejected with good parachutes.[2]

These skirmishes continued through the years, and Libya continued to support international terrorism. On 5 April, 1986, the *LaBelle* nightclub was bombed in West Berlin, resulting in American casualties. The U.S. under Reagan retaliated by bombing Libyan airports, aircraft, and facilities during Operation *El Dorado Canyon,* with much of the damage done by Air Force F-111 *Aardvark* fighter/bombers. Qadaffi's support for international terrorism began to wane after those attacks on Libya. He was getting the message!

Langford finished his time in the Navy, and eventually moved to the Florida Panhandle. He works in the construction industry during the winter months, but in the summers operates the colorful ice cream truck that cruises the beaches along the Destin/Sandestin beaches.

I just happened to purchase some ice cream bars from him for my grandkids during a trip to the beach in April, 2013, when we struck up a conversation about veterans of military service.[3]

Little did I know on first observation that he could have been a veteran on board one of our most sophisticated aircraft carriers during a time of great international strife. Most casual observers would conclude that he was just your friendly neighborhood ice cream delivery man. You just never know where you will run into an American veteran with an interesting story.

JAMES JONES

"Twenty six weeks of pure hell"

James Jones was born to Maxwell and Shirley (Daughtry) Jones in Crestview, Florida, December 16, 1964. The family eventually moved to the Onycha Community a few miles south of Opp, AL and took over operation of a store along U.S. Highway 331.

James graduated from Opp High School in 1983 and worked a short while loading trucks for H.B. Paulk Grocery Co. Looking for something more, James joined the Navy in early 1984.

Jones first attended basic training at Great Lakes Naval Training Center near Chicago in the summer of 1984. After graduation, he was sent to submarine school at New London, Connecticut. Subsequently, he was assigned to his command with COMSUB Group II where he qualified as a Torpedo-man's Mate on Tomahawk Missiles, MK-48 Torpedoes and SUBROC (an acronym for submarine rockets—either conventional or nuclear).

He later qualified as a Navy SEAL by going through 26 weeks of training at Coronado, California, just to meet a personal challenge. For anyone who has never witnessed Navy SEAL training, it is literally 26 weeks of pure hell. Even the strongest individuals—both mentally and physically—can succumb to the temptation to drop out of the course when immersed in extremely cold water for an extended period of time. Cold water is usually the proverbial "straw that breaks the camel's back," when it comes to making it through or dropping out of SEAL training.

Being the gritty individual he is, Jones stuck it out and proudly wore the SEAL insignia along with his submarine Dolphin insignia.

Jones was first assigned to SEAL Team 2 on the East Coast, and, from there, he was constantly deployed around the world. Talking to his mom, Shirley, will get you no information because she never knew where he was or what he was doing. And, James doesn't like to say too much, either. To this day, he is not at liberty to discuss details of what his outfit was doing. Reading some adventure book about Navy SEALS, or watching the national news in the last two years will reveal some idea as to the team's activities.

Wanting to be with the best among the best, Jones volunteered to become a member of the elite SEAL Team 6 (now DEVGRU). It was the DEVGRU team that went to Abbattobad, Pakistan, on May 1, 2013, and killed Osama bin Laden, the most wanted man in the world at the time.

After acceptance to ST-6, Jones traveled to Indian Head/Little Creek near Norfolk to check in with his new Team Six. Upon arrival, he was escorted to lunch, then escorted to dinner, and escorted everywhere else he went for the first day. He was told shortly to put on his dress blues uniform and appear at a hearing which, unbeknownst to him, turned out to be a court martial.

Arriving in a court in front of "more 'brass' than he could imagine," they displayed all of his records, and he thought, "Good grief, what have I done." It was very intimidating for such a young man.

What he didn't know until told later was that he was there at the beginning of a court martial for the SEAL Team Six CO, Navy Commander Richard Marcinko, who had founded the team and been its first commander, starting in 1980.

When asked how long he had been with Team Six, Jones replied, "One day." The officers on the panel told him he was dismissed and was no longer a part of Team Six, something hard to grasp, but the Navy works in mysterious ways sometimes.

After staying in his barracks for a short period, Jones returned to COMSUB Group II. In 1990, he was allowed to exit the Navy. He had been on an extension which should have taken him out to six years of service, but thanks to some legislation of Senators Gramm and Rudman, he was allowed to leave the Navy after five years of service.

Jones returned to Covington County where he now works with Robbins Electric. He enjoys deer hunting, as do so many others in and around the Onycha area.

He savors his time in service. He should, because he accomplished something very few individuals can only dream about—he completed the most rigorous and demanding training of the U.S. Navy SEALs and traveled around the world doing the things Navy SEALs do in defense of our country. We salute him for his service to the nation.

JAMES L. HOBSON, JR.

"Enemy fire was so intense that the first two aircraft commanders made go-arounds"

Major General Jim Hobson, Jr. (USAF photo)[2]

In 1977, as a young captain, I was flying with the 345th Tactical Airlift Squadron at Yokota Airbase, Japan, where I was due for another assignment. I picked up the phone and dialed the Air Force Military Personnel Center at Randolph AFB in San Antonio, Texas.

The officer on the other end of the line was then-Major James L. "Jim" Hobson, Jr., who was the Chief of Tactical Airlift Assignments. He didn't sound too enthusiastic when I promptly announced that I wanted to be reassigned to fly AC-130 Gunships at Hurlburt Field, Florida. In fact, he was downright cautious about what a drag it might be on the career of a young, upwardly-mobile, rated officer in Military Airlift Command to go fly in Special Operations. I would be "out of the mainstream," in the Air Force that existed then, and he let me know it.

Yet, I persisted, and he set me up for an assignment as requested. Little did I know it, but that would assure that our paths crossed many more times. That was the beginning of my association with one of the most distinguished and capable Air Force leaders with whom I ever flew, or for whom I served as a subordinate.

At Eglin AFB in 1986-87, we flew together to Iceland and England to visit detachments and squadrons that belonged to the 39th Aerospace Rescue and Recovery Wing—Jim was a colonel in rank and the 39th Wing Vice-Commander at the time. I was plugging along, behind the power curve as usual, and was a major by that time.

He loved to roll the dice for drinks in the Officers Club bar, I won't forget that. I also won't forget how he seemed to always be in total command while being so relaxed.

Later, at Hurlburt Field, Florida, with me in my job as Vice-Commandant of the Air Force Special Operations School, and he in his job as a Major General heading up Air Force Special Operations Command, I again served with him when I completed one of two special investigations into the cause of an AC-130 crash that occurred off the coast of Kenya. Those were good days for me, and his star was rocketing upward in meteoric speed. He had *it,* whatever *it* was.

Jim Hobson, Jr. was born in Memphis, but has an Alabama connection in that he graduated from Robert E. Lee High School in Montgomery in 1960. He then headed back to Memphis State University for an undergraduate degree in Biology. Like myself, he was commissioned a second lieutenant through Officer Training School at Lackland AFB, Texas; he received his commission in 1965.

Hobson's year of pilot training was at Williams AFB in Phoenix, Arizona, then he was on to fly *vanilla*—plain-Jane—C-130s just long enough to get the hang of flying the mighty *Hercules*. His next aircraft was the highly-specialized MC-130 in which he flew 400 hours in combat, and some 23 Top-Secret missions deep into North Vietnam. He was making quite a name for himself.

He later flew C-130J's where the aircraft snatched overhead reconnaissance satellite film as it parachuted back through the atmosphere, and before it could land in the waters of the Pacific Ocean. Shortly thereafter, he had that staff tour where he made sure I got an assignment that would land me on the most interesting and challenging missions imaginable. He did a lot of other good things along the way, but it was in *Urgent Fury*, the invasion of Grenada, 25 October, 1983, that he showed his mettle.

Hobson was by then the 8th Special Operations Squadron Commander, and he was flying one of the MC-130s in the flow to drop U.S. Army Rangers on Point Salines Airfield during a time of heavy enemy anti-aircraft fire.

Deadly AAA weapons such as the Soviet-built ZU-23-2 were taking aim at the MC-130s. Enemy fire was so intense that the first two aircraft commanders made go-arounds, leaving Hobson, as number three, the one who had to make something happen.

Hobson decreased his fly-by altitude to 400 feet above the ground, which is extremely low for a parachutist who would get only one good swing and hit the ground. His Ranger troops were able to get on the ground, seize the airfield, and proceed as planned. Without his grit and determination, the operation's success would have been forestalled.

For his valor and leadership under fire, Hobson was named recipient of the Air Force's *Mackay Trophy*, an award given for the most meritorious flight of the year for 1983.[1]

Various other command positions would follow in rapid succession, including one as Vice Commander of the 39th ARRW at Eglin, where we flew together, some trips to England. He ultimately became commander of the Air Force Special Operations Command where he had 11,900 personnel responsible to him, and with some 100 aircraft at his disposal, to carry out a world-wide special operations air mission, including some in the counter-terrorism realm.

When the U.S. Air Force dropped our first troops in Afghanistan after the 9/11 terror attacks, most of those same aircraft would be in play, and the aircrews flying them would be trained under the tactics, techniques and procedures that General Hobson and his AFSOC leaders had established as standards to be met.

After a stellar career, and one in which I was fortunate to know and work with him, Hobson retired from the Air Force on August 1, 1997. But, even then, he and wife Diane (Jetton) weren't finished serving their country. There was no time to rest on one's laurels.

Hobson's Air Force time was followed by his 12 years of service as President and CEO of Air Transport International, in Little Rock, Arkansas.

During that time, he was honored at Maxwell AFB's Air Command and Staff College in what's termed a *Gathering of Eagles*, where those such as Hobson share their experiences with America's military leaders of the future.

Previous *Eagles* to be so honored included those such as Jimmy Doolittle, Curtis Lemay, Gene Kranz (of NASA), Paul Tibbets, Jr., who flew the *Enola Gay*, Greg "Pappy" Boyington—remember the TV show *Baa Baa Black Sheep?*—Johnny Alison, "Chuck" Yeager, Charles "Chick" Cleveland (Korean War fighter Ace), and so many others of that stature.[3]

What a terrific leader and aviator he was, and I am one of thousands who served with and for him who would stand in line on a rainy day to salute Major General (Ret.) James L. Hobson, Jr., for his so-numerous contributions to our nation's national defense. He truly is a Great American!

PERSIAN GULF WARS

In the late 1980s, tension mounted between Iraq and Kuwait over several issues. Generally speaking, because of a glut, Kuwait was willing to decrease oil prices on the international market, and Iraq was not, claiming that for every US $1 per barrel drop in oil prices, Iraq was losing US $1 billion in revenue. Iraq already owed Kuwait US $14 billion from borrowing to finance its war with Iran.[1]

Iraq also owed Saudi Arabia, to the tune of about $16 billion, with no way to repay the debt with oil prices dropping. "Saddam (Hussein) requested cancellation of the loans; he was refused. Next, he requested additional credits to help rebuild the shattered economy (following the 8-year war with Iran)... His requests were not honored. Finally, Saddam requested that the price of oil be raised and production reduced as a way to maximize Iraqi oil revenue. Not only was the request denied, but Kuwait engaged in 'slant drilling' from Kuwaiti soil into the Rumalia oil fields claimed by Iraq, in effect stealing Iraqi oil (Kuwait denied the claim)."[2]

The U.S. had re-opened relations with Iraq in 1985, believing he was a better alternative than the Ayatollahs in Iran. Despite Saddam's threats toward Kuwait, the U.S. pretty much ignored the possibility that he would invade.

"Two weeks before the invasion, April Glaspie, US Ambassador to Iraq, assured Saddam that the US had no interest in oil negotiations among the Arabs or in boundary disputes (added note: Iraq had historically claimed Kuwait as an Iraqi province). When asked after the war why we did not foresee the impending invasion, she replied, 'We didn't believe he was that stupid.'"[3]

On 2 Aug 1990, Iraq invaded Kuwait and took over the entire country, precipitating the US response that became known as DESERT STORM. The US would kick Iraq out of Kuwait, but stopped short of going all the way to Baghdad to comply with a UN mandate that only called for the ejection of Iraqi forces. After the 9/11/01 terror attacks, the US went back a second time in March, 2003, occupied Iraq, captured Saddam Hussein, and turned him over to a new Iraqi government. Some of the stories that follow involved veterans who participated in those operations.

MICHAEL P. AND CHERYL L. HYNES

Michael P. Hynes in uniform

Michael Patrick Hynes was born in March, 1959, in Syracuse, New York. His father and mother were originally from the small town of North Adams, MA, but had moved to Syracuse just before Michael was born.

At age ten, his parents moved back to North Adams. Mike's family was considered middle class. His mother, Mary Lou, was a stay-at-home mom, and his dad, Richard, worked at most anything to make ends meet.

Mike was industrious as a youngster. He borrowed sixty dollars from his grandfather to buy a newspaper route during high school. He also made money working at a gas station pumping gas and doing small repairs for the station. He always wanted to have some money in his pocket, and the work ethic that was handed down from his parents gave him the ability to earn spending money. More importantly, it gave Mike the perspective on how important it was to work for what you wanted. It would serve him well throughout life.

Hynes graduated from Drury High School in 1977 and tried his hand at college away from home, but the lure of having a full-time job that paid fairly well at North Adams' premier employer, Sprague Electric, meant that Mike would not be going back to college, at least not full time, though he knew that the key to long-term prosperity was a college education. He continued his education through educational assistance provided by his employer at North Adams State Teachers College.

Although he was working full time and going to school, deep down, he knew he was going nowhere fast in North Adams. While scoring a summer basketball game for the city, Hynes got a lucky break.

Sergeant First Class (SFC) Robinson was the area's Army recruiter, and he also coached one of the girls' basketball teams. He and Mike had friendly conversations before and after the games.

Then in the late fall of 1979, the Iranian Hostage Crisis would provide the motivation for Mike to venture down to SFC Robinson's office in the North Adams Post Office to see what kind of opportunities the Army held.

As Mike stepped into the Sergeant's office, Robinson looked up, saw him and asked, "What took you so long?"

Private Hynes' first day of Army active duty was 16 May, 1980. He enlisted as an Intelligence Analyst and received Basic Training at Fort Leonard Wood, MO, then Advanced Individual Training at Fort Huachuca, AZ.

As an enlisted soldier he served first at Fort Meade, MD, then an overseas assignment to Korea and back to Fort Huachuca. He continued his college education at night, earning his Associates degree while in Korea. As Mike gained more experience, he began to aspire to be an officer. His chance came after re-enlisting for Airborne School and a reassignment to Fort Bragg, NC.

While at Bragg, Mike learned of an R.O.T.C. Scholarship program for active duty soldiers. The Army had 75 two-year scholarships Army wide. Although he thought it was a long shot, he applied for what he considered a golden opportunity. After not hearing anything from the scholarship board, Hynes decided to give the committee a call to find out his fate.

As luck would have it, the Army didn't have enough two year R.O.T.C. applicants to fill all 75 available scholarships, so all that had applied that year would attend school the next fall. Mike started his junior year at Campbell University, located just north of Fort Bragg.

In the spring of 1987, Second Lieutenant Hynes graduated as a newly-minted Signal Corps officer and headed to the basic course, then on to Fort Bragg, N.C., where he knew life would get interesting, given the future events that Operation DESERT STORM would entail.

He performed the typical lieutenant duties with his company until one day the Battalion XO, Major Sam Higdon, asked him if he would like to transfer to the battalion across post that Higdon had just been selected to command.

The 112th Signal Battalion (Special Operations)(Airborne) was a relatively new battalion that was stood up to support the Special Operations community with long-haul communications. Mike jumped at the chance to be the battalion's logistics officer. 1LT Hynes assumed the duties as the 112th Signal Battalion Logistics Officer in August, 1990, the same time the US Army was finishing up a huge exercise called "Internal Look" which was used as the preliminary rehearsal to Operation DESERT STORM.

The battalion had never deployed to a war zone, so the standard operating procedures for deployment were not fully developed. This meant LT Hynes had to do some improvising if he was to get the battalion uploaded and ready for upcoming combat operations.

Although Fort Bragg had plenty of experience in deploying units, the 112th Signal Battalion had none. There were vehicles that needed to be painted, rations to be ordered, and soldiers to be prepared, all in just thirty days to "wheels up." Through teamwork, the battalion was ready to go and LT Hynes was on the first plane into King Fahd International Airport, Saudi Arabia.

Working around the clock, LT Hynes insured all follow-on aircraft arrived to the theater and were ready to convoy to the Battalion's staging area at King Khalid Military City. Hynes assembled the unit and mapped out the route to the City and told his Battalion commander he would see him in a few days.

LTC Higdon was perplexed as to why the lieutenant was not moving with the rest of the battalion. LT Hynes was aware that pallets of material were easily misplaced or stolen and decided that he would remain behind to insure the pallets of the battalion's important gear would make it in one piece. The commander thought about it and said goodbye to his log officer. Several days later, after making coordination for the follow-on equipment, he joined the rest of his unit, accompanied with the equipment and his tactical vehicle in a C-130 aircraft.

For the rest of the deployment, Hynes oversaw the logistical requirements of the battalion and, at the request of Higdon, Mike doubled as personnel officer. This meant that not only was he responsible for getting the battalion back to Fort Bragg, he was also responsible for all of the personnel actions that are required of soldiers in a war zone. Hynes was awarded the Bronze Star for his efforts during Operation Desert Storm.

From Fort Bragg, he went on to serve at the Joint Task Force level and became the Operations Officer for the 124th Signal Battalion, 4th Infantry Division. The 4th ID was charged to complete the *proof of concept* for the Army's initiative to digitize all Division and Brigade combat units.

Based on his experience at the 4ID, Mike was transferred to the Pentagon just in time to lead the effort to equip the Army's deploying units to Iraq and Afghanistan. He also had the distinction to serve General Benjamin S. Griffin as his Aide de Camp at Army Material Command, not the first time that COL Hynes had the pleasure to work for Gen. Griffin. Throughout his career, Mike had worked for GEN Griffin no less than five times and says that "my association with GEN Griffin is one of those lucky breaks that contributed immensely to my success."

Mike Hynes ended a 31-year career in the Amy in November of 2012. It had been a long and challenging trek from his days as a private to the rank of colonel. He commanded at both the company and battalion levels, and he served throughout the United States, Alaska, and Korea.

He now makes his home in Northern Virginia with his wife, Cheryl, and their two children, Josh and Carrie.

Mike's wife, Cheryl L. Hynes, also served with distinction as an Army Signal Corps officer where, as a captain, she commanded Alpha Company, 124th Signal Battalion, 4th Infantry Division, at Fort Hood, Texas.

An Auburn University graduate, she enlisted in the Army Reserves in 1991 and started out as a specialist (E-4) right out of Fort Jackson's basic training, subsequently picking up a skill as a 31C MOS, Radio Operator. She installed, operated, and performed unit-level maintenance on AM and TACSAT (man-portable) radios, as well as high-frequency data systems, among many other duties. She went active duty in 1992.

Cheryl served a 19-month tour in Germany, then three months at Fort McClellan before starting OCS at Fort Benning. Her father, this author, had the privilege of administering the commissioning oath of office at Fort Benning on May 20, 1994.

After being commissioned, she attended Airborne training at Fort Benning, accomplishing the required five jumps to earn the Parachutist Badge. When her parents asked her why she wanted to go to jump school, she replied, "I want to be competitive for promotions with all the men in the Army," and she was!

She was a Platoon Leader and XO in a signal company, a Battalion S6 in air defense artillery (as well as in a Patriot Missile battalion), and served in the 2d of the 43d Air Defense Artillery Battalion, XVIII Airborne Corps. She also was Battalion S1, 13th Signal BN, 1st Cavalry Division, proudly wearing the Cavalry spurs that she worked so hard to acquire.

After assuming command of the company at Fort Hood, just like all other Army soldiers, she spent time in the field at Fort Irwin National Training Center, roughing it in a sleeping bag on the desert floor alongside her *Humvee* during training iterations. She had 137 soldiers reporting to her and was quite the disciplinarian.

Cheryl eventually moved to a different field, going to school at Redstone Arsenal, cross-training into Army Acquisition, and then working and managing Army Enterprise IT Systems, to include Army Knowledge Online (Portal).

She earned two Masters of Science Degrees (in Management and Information Systems Management) along the way and rose to the rank of major, but eventually decided to leave active duty and enter Civil Service in the Acquisition field. At present, in Washington, D.C., she is an Enterprise Services IT Portfolio Manager, managing Army service requirements and investments.

Retired Colonel Michael Hynes continues to serve his country, and is now working as a civilian in the Office of the Deputy Assistant Secretary of Defense (OSD), Maintenance Policy and Programs, where he is Director, Maintenance Integration and Readiness Assessment.

Mike earned a Bachelor's degree and three Masters Degrees while on active duty and routinely speaks at Veterans' Day activities in his hometown of North Adams.

Cheryl and Mike Hynes lead busy lives, with both juggling work as civil servants, fighting traffic on and near Washington, D.C. freeways, and raising two youngsters through adolescence.

Mike and Cheryl Hynes are dedicated individuals, and both are to be commended for devoting the major part of their lives to the defense of our country. Their families, friends and work partners salute them for their efforts.

PATRICK BRIAN GURIS

"Aboard the Lincoln when the U.S. Navy captured some of the Somali pirates"

Promotion to Master Chief Petty Officer (Pat, Jr. left, and Pat, Sr. right)

The U.S. Navy created the rank of Chief Petty Officer some 120 years ago, specifically designating the CPO rank for those who are capable of being "a leader of sailors." One of those who would eventually fill that rank is Master Chief Petty Officer Patrick Brian "Pat" Guris who retires in March, 2014, after 29-plus years of active duty service.

Pat Guris was born May 28, 1965, in Ilion, N.Y., to Jim and Barbara (Mills) Guris. Pat was the middle child, sandwiched between sisters Kelly and Mary. His father, Jim, had grown up in Herkimer, about 72 miles west of Albany, and had also been in the Navy. So had several other ancestors.

Ilion is a small town in upper, mid-state New York known primarily for the 194-year-old Ilion Firearms Plant, the Mohawk Valley's largest employer and the nation's longest continuous manufacturer of firearms.

Pat's dad, Jim, took a job that landed him in the Florida Panhandle in a neighborhood where they lived two houses down. As a result, Pat completed much of his schooling in the Ft. Walton Beach, Florida area.

Some elementary grades were at St. Mary's. Grades 7, 8, and 9 were at Bruner Junior High before he moved on to Fort Walton Beach High School. There, he was an offensive lineman on the Vikings football team for the years 1981-83. While still a student, Pat worked at the "Fish House Restaurant" in Destin, bussing tables and doing all the other things required to help make a business a success.

In 1984, Pat joined the U.S. Navy, continuing in the tradition of his father, Jim. He wanted AMS Structures for a career path in the Navy, but his recruiter informed him there would be a considerable wait to enlist, so he chose Engines that allowed him to get in the Navy the day after Christmas that year.

He was first off to basic training at Orlando, then up to Memphis to complete his technical training. In the latter part of 1985, he was assigned to Pensacola Naval Air Station with a helicopter squadron to work on engines as an Aviation Machinist Mate which includes quite a great responsibility, as follows:

"...maintain aircraft engines and their related systems, including the induction, cooling, fuel, oil, compression, combustion, turbine, gas turbine compressor, exhaust and propeller systems; preflight aircraft; conduct periodic inspections on engines and engine-related systems; field-test and adjust engine components, including fuel controls, pumps, valves and regulators; remove, repair and replace compressor and turbine blades and combustion chamber liners; preserve and de-preserve engines, engine accessories and components; and supervise engine work centers."[1]

In December, 1989, Guris was re-assigned with an A-6 *Intruder* squadron at Oceana NAS in Norfolk. It was there he had his first deployment aboard the Navy carrier USS *Saratoga*, a Forrestal-class super carrier which was the second Navy carrier named for the Revolutionary War battle. The *Saratoga* was commissioned in 1956 and was decommissioned in 1994. Pat wore the *green vest* on deck. Ordnance handlers wore red vests, the quality control folks wore white, the airplane handlers wore yellow, and refuelers wore purple.

For those who have never been on an aircraft carrier, the noise is deafening, and hand signals are used for all movements and activities by those who wear the different colored vests. The action might be described as being akin to an orchestra, with the carrier "air boss" ruling the roost, and different performers continually working in and out of the aircraft preparation, launch and recovery activities.

The *Saratoga* was deployed in the Red Sea August '90 to March '91 in the First Gulf War when the Bush 41 Administration worked with a United Nations Coalition to remove Saddam Hussein's Iraqi forces from Kuwait.

The ship launched many flights in support of operations, including that of Scott Speicher, correctly assumed to be the first American casualty of the war.

Another *Saratoga* aircraft shot down was an A-6E *Intruder*. Bombardier/Navigator LT Jeffrey Zaun (with fellow crewman Wetzel) was paraded before cameras by his Iraqi captors, but was eventually returned to American forces and was able to meet the *Saratoga* when it returned home.

Saratoga-based U.S. Navy SEALS conducted the first wartime boardings of merchant shipping in the Red Sea in support of Operation Desert Shield.[2]

According to Guris, one of their worst disasters occurred when sailors were being ferried from the *Saratoga* to Haifa, Israel, during rough seas in December, 1990. The ferry capsized, and some 21 sailors were lost, something Guris found tough to fathom.

The USS *America* and USS *John F. Kennedy* carriers were also there in the Red Sea. Our Navy had three other carriers deployed in the Indian Ocean in one of the most concentrated incidences of U.S. naval power in modern times. Guris had another deployment on the *Saratoga* in the Mediterranean in 1992-93, as well.

It was during this time in his first six years of duty that Pat Guris was promoted to E-6, as he was proving himself to be an outstanding Aviation Machinist Mate and one who could get things done in a professional and timely manner.

After his time on the *Saratoga*, he was assigned shore duty for two years at Virginia Beach (1994-96). Then, he was on to Jacksonville until Cecil Field was closed. A move to Beaufort Marine Corps Air Station followed where he served for two years.

From Beaufort MCAS, he was part of two deployments, one on the carrier USS *John F. Kennedy*, and one on the USS *George Washington*. His most recent deployment was on the USS *Abraham Lincoln*.

One of the more interesting events while aboard the *Lincoln* occurred when the U.S. Navy captured some of the Somali pirates that had been grabbing ships and crews near the Horn of Africa. The raiders were brought aboard the *Lincoln* and locked up in the ship's brig. When the ship pulled into shore, Pat took his dad down to show him where the pirates were kept, as a point of interest.

Pat was assigned shore duty in March, 2011, so he missed the important event of May 1, 2011, when U.S. Navy SEALS known as DEVGRU eliminated the world's most wanted terrorist Osama bin Laden in Pakistan. However, other changes of significance pulled him into the heart of the action.

On July 14, 2011, The Department of Defense's first F-35 *Lightning II* joint strike fighter touched down at its new home at Eglin AFB, Florida, marking a major milestone in the nation's military history. Upon arrival, the jet officially became part of the Air Force inventory and now belongs to the JSF training unit, the 33rd Fighter Wing.[3]

As the first of its kind in the Department of Defense, the wing is responsible for F-35 A/B/C *Lightning II* pilot and maintainer training for the Marine Corps, the Navy, the Air Force, and, in the future, at least eight coalition partners. Initially, 59 aircraft and three flying squadrons, one for each service/aircraft variant, will be established at Eglin.[4]

With all the changes involved in bedding down the new F-35 at Eglin, then-Senior Chief Pat Guris had suddenly found himself reassigned to Eglin AFB in November 2008 as part of an advanced team where he was working with Air Force, Marine, Navy, and civilians in assuring preparatory actions were taken to ensure a smooth transition to Eglin. With the assignment was an implicit understanding that no promotion could come from the assignment. With that in mind, Guris put in his papers to retire once his time was up.

The Navy, however, wasn't about to let a good man go. By this time, all the deployments and absences had taken a toll on Pat's first marriage, and he had divorced and been a single parent of two girls and a boy for much of the time. By the time of the assignment to Eglin, he had remarried to present wife, Julie.

On a day off from work, he and dad, Jim, were remodeling a house Pat and wife, Julie, had bought in Navarre when an incoming call requested his presence at Eglin ASAP (saying he should check the promotion list). Pat was confident he wasn't being promoted, so he brushed off the call.

Pat reminded the caller he was off that day and had just removed the front double-door of his home for replacement, and said he couldn't leave the job with no door on his home.

Shortly, another call came from a different military caller from Eglin. Suspecting nothing, Pat was abrupt in telling the caller he was tied up. About 30 minutes later, a third call came, this time one from Norfolk, with the news that he had been selected for promotion to Master Chief Petty Officer. Suddenly, his dad, Jim, overhead several, "Yes sir, Yes sir, I'll get on that right away" responses.

Suddenly, Pat took off from his now less-than-urgent remodeling job and rushed over to Pensacola to purchase all-new Master Chief insignia, since the insignia was not available at Eglin. It didn't take long for Guris to pull those retirement papers either, and then input a new projected retirement date of March 31, 2014, which will be 29 years and 3 months to the day from when he joined the Navy.

After making E-9, Pat had two more assignments, including one with Wing Command of Air Wing Two which deployed again for 6 months on the carrier *Lincoln* during the period September, 2010 to March, 2011. That deployment was followed by another from December, 2011 to August, 2012. He's now in Lemoore, California on his terminal assignment.

Pat's children are Ashley, living in Ohio; Patrick Jr., living in Pensacola, and Jessica who lives with Pat and Julie in Lemoore. Julie also has a son, Mathew, who is with the family. Despite the hardships of his many deployments and absences, they all have bright futures.

Pat said the best things about being in the Navy were serving the country, the friendships he made, the comradeships developed, and esprit de corps. He will miss those aspects of Navy life. The worst part was being away from home so much and the frequent re-locations. He won't miss those aspects.

For the future, he plans to move back to Florida and hopes to work in the same line of work, maybe with F-35 programs, or something at Whiting field with naval aviation.[5]

Pat's parents, Jim and Barb, plus all the friends and neighbors who surrounded him in his youth, are very proud of Pat Guris for the outstanding service he performed for our country in its time of need.[6]

It is men and women like Pat Guris who sacrifice so much in the way of long deployments, missed birthdays and holidays, and numerous other personal deprivations of the many things in life that so many take for granted who assure we maintain the freedoms we have in this country.

I know all of America joins me in saluting retiring Master Chief Petty Officer Patrick Brian Guris for his outstanding 29-plus years of service to our nation. Well done, Pat!

BOB AND GUY BUICE

"Guy served seven years with the 82nd Airborne Division at Fort Bragg, completed tours in Iraq and Afghanistan, and is now stationed with NATO HQ"

Robert J. "Bob" Buice was born to Len and Mary (Roberts) Buice on May 28, 1934, in Buford, Georgia. Bob's father worked as a mail clerk on the railroad where he sorted mail as he traveled aboard the train. Bob's mother died when he was 15, and, with his father away 3-4 days a week, he and his sister often made do, with a lot of help from neighbors watching out for them and assisting.

Buice attended Buford High School, then completed one year at North Georgia State Military School before transferring to Georgia Tech where he majored in Industrial Engineering. His last two years at Tech included working on the C-130 *Hercules* at Lockheed Corporation in Marietta.

Bob volunteered at Atlanta's Peachtree Street Marine Corps recruiting station in 1954, but was rejected because he weighed only 130 pounds. After waiting eight months, he tried again and was then accepted.

After completing OCS, Bob found himself in interesting, exotic locations shortly thereafter. Egypt's Prime Minister Colonel Gamal Abdul Nassar seized the Suez Canal from the British in July, 1956, and then announced Egyptian nationalization of the Canal and its operating Suez Canal Company. This ostensibly was in retaliation for the reneging of an agreement by the American and British Governments to finance construction of the Aswan Dam.

The Suez Canal represented the main source of supply of oil for Britain and France, and the potential loss of those supplies represented an economic threat that they could ill ignore.

In this crisis, Bob as a second lieutenant found himself a rifle company commander on the way to Port Lyautey NAS, Morocco, about 75 miles northeast of Casablanca, in a deployment with some 200 Marines. They flew over on three slow-moving Marine C-119 *Flying Boxcar* transports, with the trip taking two days. All the pilots were enlisted gunnery sergeants, with officer co-pilots.

A Colonel Silvie was in charge in Morocco (Silvie had been commander of the Parris Island training center when some eight Marine recruits died in a swamp there in 1955).

The Marines were deployed to safeguard an antennae farm about 20 miles outside Lyautey, which was nothing much more than a crossroads location, but was an extremely important facility. All messages from Europe were transmitted through the antennae farm, so protecting message traffic in a crisis was very important. There were also some warheads stored at Port Lyautey.

At the base, the French Foreign Legion protected the outside of the base, and Bob's fellow U.S. Marines covered the inside. He remembered the Legion troops were some of the nastiest he had seen and "smelled to high heaven."

They also had a pet ferret that they stuck down in rabbit holes, with the result that the ferret grabbed and pulled out the rabbits for the troops to eat. They pretty much lived off the land.

They also had little standardization as to weapons. One had a Belgian rifle, another carried an M-1, etc. As long as they had bread and wine, however, they seemed fine.

He recalled the Base CO kept a water-cooled machine gun pointed at the Legion troops. He didn't trust them one bit.

Of the 200 Marines there, 23 did not have a high school diploma. Buice set up classes, and, by the time the Marines came back Stateside, all 23 had diplomas, something Bob is still very proud of today.

On another occasion in 1958, he was part of a deployment to Beirut. Seems the Russians were playing games on the Turkish-Bulgarian border, and the U.S. needed a response. There was a mile-long runway and they had responsibility for the terminal and airport.

Off one end of the runway was a ghetto with gunfire often coming from that area. They discovered rifle fire was coming through a stove pipe so as to avoid discovery. On the other end of the runway was an olive orchard, something very expensive, and enemy fire came from the orchard quite often as well.

A patrol leader got tired of the incessant sniping, so, on one very dark night, he lit off a flame thrower which massively illuminated the entire area. They found two burned up rifles and a burned out olive orchard. The U.S. government had to pay a hefty sum as compensation for the owner's loss of the orchard. The orchard problem made the news big time on the Huntley-Brinkley NBC Evening News report.

Along the way, Buice married his childhood sweetheart, Lou Wages of Buford, and they had a son, Guy, now an Army officer, as well as a daughter, Leslie, who is a therapist.

In his 5 years of active and 23 years of reserves service, Bob also served as Adjutant to a Marine Fighter Squadron at NAS Atlanta where he did a lot of the paperwork that pilots abhorred. He also was CO of a Logistics Branch of the Marines Reserves.

As Adjutant, he became involved in the paperwork of aircraft crashes that took several lives. Two pilots in one aircraft crashed off the end of the Atlanta runway. Their bodies were found still strapped in their seats up against a pine tree in a trailer park.

Another incident, off the end of the runway at Beaufort, S.C., occurred when the pilot was attacking a towed target, became disoriented, and crashed because of the vertigo. The crash was so horrendous that the investigators found only a portion of the palm of the pilot's *glove*. Nothing else was identifiable.

A third incident involved a Marine colonel flying an aircraft departing NAS Atlanta (across the runway from the Dobbins AFB side) who was killed when his aircraft wing folded up and rolled over into the ground. Pilots did the technical things such as identifying bodies and the investigating, and Buice then wrote up the reports, so he saw a lot of gruesome details.

In his civilian career, Bob Buice was a District Sales Manager with Allstate Insurance Co. He now lives on Lake Martin near Alexander City, AL.

Bob's son, Guy Buice, went to Jacksonville State University in Anniston on an athletic scholarship. Some of his buddies wanted extra credits, so they convinced Guy to talk to the colonel that ran ROTC.

Guy did, and he enrolled in ROTC, also on a military scholarship. But first, he had to get three medical waivers to enter military duty. He had asthma, his eyes were bad, and he lacked chest-muscle mass. Guy cleared the waiver hurdles by doing a certain number of push-ups.

Guy graduated in 1993 as a second lieutenant and went to Fort Huachuca, Arizona, for training as an Intelligence Officer. In college, Guy had a minor in Germanic Language, and he desired to have a tour in Germany. To assure that tour, he first volunteered for a tour in South Korea near the DMZ, where he got a bad case of frostbite, but ultimately succeeded in getting that badly-wanted tour in Germany.

Guy Buice attended required service schools such as Command and General Staff College, served seven years with the 82nd Airborne Division at Fort Bragg, completed tours in Iraq and Afghanistan, and is now stationed with NATO Headquarters, in Mons, Belgium.

When he received his last promotion to lieutenant colonel, Guy convinced his dad to fly to Belgium to administer the oath of higher office, something that was a high honor for both. Guy had leased the Officers' Club, sponsored an open bar, and entertained representatives from 18 different countries in attendance.

Guy Buice now has over 20 years of service. Combine that with his father's 28 years, and it won't be long before they reach a combined half-century of service to their beloved USA. What began in 1955 with Bob's service continues today in 2013 and may go on for a much longer time.[1]

What patriots they are, and we are thankful there are family combinations such as the Buices that serve our country in such an outstanding fashion. Hooaah!

IRAQ/AFGHANISTAN

Pentagon Burning after crash of American Airlines Flt 11 on 9/11/2001

The wars in Iraq and Afghanistan were precipitated by the hi-jacking of four commercial airliners by Arab terrorists and the subsequent attacks on the World Trade Center, the Pentagon, and one of the hi-jacked aircraft that crashed in a field near Shanksville, Pennsylvania, on September 11, 2001.

Nearly 3,000 Americans died on that day, a day of carnage that rivaled and even exceeded the Japanese sneak attack on Pearl Harbor that precipitated America's entry into World War II. The political, military, and economic pillars of our country were all attacked with great success.

It's easy to make the argument that we are in fact in another World War at the moment, as we fight against a very well-organized international terrorism phenomenon.

What started as a localized group of al Qaeda terrorist training camps in Afghanistan has morphed into a much wider network now including growing threats in Africa and the Middle East. Mali, Somalia, and Yemen are just some examples.

For decades during the 20th Century's Cold War, American and NATO planners spent enormous resources to be able to repel a Soviet-Warsaw Pact armored attack through the *Fulda Gap* and lowlands in Europe which would have quickly placed West Germany's financial center of Frankfurt in Soviet hands.

With the fall of the Iron Curtain in 1989, the reunification of Germany in 1990, and the subsequent withdrawal of Soviet forces, the Fulda Gap lost its former strategic importance, yet remains a powerful symbol of the Cold War.

With the *change in the nature of warfare* in the last half of the 20th Century, many new "Fulda Gaps" have emerged. They are far different, and more varied, but must be dealt with.

They may be in the form of a tiny fishing skiff loaded with explosives in a harbor in Yemen that can severely damage a U.S. Navy ship and kill dozens.

They may be in the air terminals of Portland, Maine, or Logan or Newark Airports, where those with box-cutters are not detected and prevented from boarding, as happened on 9/11/2001.

Or, they may be in the offices of a Saudi Arabian travel agency that issues visas to the U.S. with inadequate screening as to the dangers of potential terrorism the traveler poses.

They may be in the personage of those recruited to our armed forces because of their language abilities, but not watched closely enough to detect when they are on the brink of carrying out *jihad*, as happened at Fort Hood.

They are at the border entry stations at El Paso and Laredo, or at the U.S./Canadian Bridges at Niagara Falls or Detroit. They are the narrow passageways/alleyways next to American Embassies abroad which offer perfect opportunities for sophisticated truck bombs to level a building and kill hundreds, as happened in Nairobi, Kenya in 1998. They are at the offices of rental truck companies or private aviation training companies in the United States.

They may be on the tarmac of airports such as Frankfort or other overseas terminals. For example, on March 2, 2011, an Air Force bus loading airmen to send to Afghanistan via Ramstein was unexpectedly attacked by a stranger, Arid Uka, who asked whether the troops were headed to

Afghanistan. When Senior Airman Nicholas Alden said yes, Uka pulled out a pistol and shot and killed Alden. He then climbed aboard the bus and killed the driver, Airman 1st Class Zachary Cuddeback. He wounded two other airmen before fleeing. Though he was caught and sentenced to life in prison, the fact remains that there are dangers everywhere for our military personnel.

The gaps may even be in the hardware stores where those who hate Americans and our way of life purchase pressure cookers, pack them with nails and ball bearings, and detonate them at the end of public events such as the Boston Marathon, as so recently happened.

These modern-day Fulda Gaps are so numerous that our country could die a death of a thousand cuts, if not considered when devising defensive strategies.

These gaps are here, and our government and military services must recognize that a sea change has occurred and strategies, doctrines, equipment, and training must be tailored to counter those threats, something that not always happened adequately in my days in service.

To understand fully the contributions of those in the Iraq or Afghanistan Wars, one must understand the background and the precipitating forces that are rooted in international terrorism. The subject is broad, and certainly too comprehensive for this author to discuss completely in this short document. What can be covered are instances of heroism by those individuals such as Marine Corporal Jason Dunham.

CORPORAL JASON L. DUNHAM

"...a hand grenade fell from the Iraqi's grasp"

Corporal Jason Dunham's Recruit Photo

The town of Florala, Alabama, is in the deep South. The town of Scio, New York, is in the upper reaches of the North. The Census population of Florala in 2000 was 1,964. The population of Scio in 2000 was 1,833. Florala residents enjoy the small-town benefits of most residents knowing most others. Scio residents enjoy the same benefits. Scio has one stop light, Florala only has four.

There is another thing they have in common. Both tiny towns have Medal of Honor recipients whose families received the recognition for their heroes *posthumously*.

And, in the case of 22-year-old Marine Corporal Jason L. Dunham of Scio, his actions of heroism were almost identical to those of 21-year-old Army Sergeant Rodney J. Evans of Florala. They both deliberately exposed themselves to almost certain death in order to save the lives of their fellow troops. Both paid the ultimate sacrifice, and their fellow men in uniform lived to fight another day.

Corporal Jason Dunham was a squad leader in Fourth Platoon, Kilo Company, where he was responsible for nine other Marines (Kilo was part of the Third Battalion, Seventh Marine Regiment).

On April 14, 2004, Dunham and his squad were on patrol in Husaybah, Iraq, in the Al Anbar Province, near a portion of the Iraqi-Syrian border, when they rushed to help other Marines caught in a cross fire by heading toward the spot from which they thought the fire was originating. On the way, near an intersection, they attempted to search a Land Rover vehicle and its Iraqi occupant.

At first glance, they had spotted the muzzle and wood front grip of an AK-47 rifle poking out from under the floor mat, so the vehicle and occupant were highly suspect. As they announced their intentions to search, an Iraqi in the vehicle lunged suddenly and grabbed Dunham around the neck in an attempt to kill him.

In the life and death struggle that ensued, a hand grenade fell from the Iraqi's grasp and fell in plain view of Dunham, exposing him and his troops to the extreme danger of an exploding hand grenade. A hand grenade with pin pulled and safety handle released has about 3-5 seconds before exploding with enormous destructive power.

Having thought through this scenario before, and having discussed the hypothetical of it on numerous occasions, Dunham within a split second removed and placed his own helmet over the grenade, then threw himself on top of and wrapped his arms around the helmet, thinking and hoping against all hope the helmet would be sufficient to protect himself and his fellow Marines.

The British-made *Mills Bomb,* a classic pineapple-shaped hand grenade, exploded while Dunham covered it. The explosion released fragments that tore through the helmet with ease and cut a swath with deadly debris in its path, striking several of Dunham's fellow Marines who were rushing to his side to help subdue the Iraqi. He was credited with saving the lives of at least two of his fellow Marines while giving his own.[1]

Michael M. Phillips, a staff reporter with the *Wall Street Journal,* was with Dunham's battalion and later wrote a book about Dunham, including passages regarding the seriousness of his injuries once he was airlifted out on a helicopter and triaged:

"Dr. Hessell searched carefully through the blood and pulp on the side of Jason's head and guessed that he had been hit by three grenade fragments. One, he thought, had penetrated the left side of his skull not far behind his

eye. The second had entered the brain slightly higher and farther toward the back of his head. The third had punctured his neck, but didn't go in very far. Jason's heart rate and blood pressure were stable, but he had begun 'posturing'—an almost fetal curling of the arms and clenching of the fists that signalled serious injury to the brain."[2]

After Jason endured several days suffering with his mortal wounds, his parents agreed to remove all life support, and Corporal Jason Dunham died eight minutes afterward, at 4:43 p.m. on 22 April, 2004, at Bethesda Naval Hospital.

The Commandant of the Marine Corps had rushed to his bedside to join his parents, Deb and Dan Dunham, and present a Purple Heart only a few minutes before he drew his last breath. Some of his fellow Marines were seriously injured from the attack, but they all ultimately survived, thanks to his personal sacrifice.

Dunham's funeral was on May 1, 2004, in his high school gymnasium. Of Scio's 1,833 residents, more than 1,500 were in the gymnasium to bid goodbye to their hometown hero who had been a great athlete at the school and one of the most popular students ever to graduate.[3]

Jason L. Dunham was buried in Fairlawn Cemetery, and Michael Phillips described his funeral proceedings:

"Seven Marine riflemen stood near the gravesite. They fired three salvos, and a bugler played taps. The Marines folded the flag that had been on Jason's coffin into a crisp triangle and handed it to Dan. The Marines saved three empty shells, one to be tucked into each corner of the folded flag."[4]

Boyhood friends of Rodney J. Evans had been present for an almost identical military ceremony for Evans in Florala, Alabama, on a hot summer day in 1969. That funeral pretty much mirrors the description of the ceremony for Dunham in that formal military burial honors were bestowed.

In the East Room of the White House, President George W. Bush on 11 Jan 2007 presented a posthumous Medal of Honor to Jason's parents, Deb and Dan, with Jason's sister and brothers in attendance.

On 20 March 2007, the Navy reported that a new *Arleigh-Burke Class* guided missile destroyer would be named the USS *Jason Dunham* (DDG-109).

The Post Office in Scio now bears Jason Dunham's name.

Dunham was the first recipient of the Medal of Honor of any American in the Iraq War, and the first to be presented since the end of the Vietnam War.[5]

My hometown of Opp is 20 miles north of Florala. When my family and I drive south toward Fort Walton Beach along U.S. Highway 331, about two miles north of downtown Florala, we always look to the east and take note of the sign that denotes Rodney J. Evans Drive.

One can't help but wonder what it is inside men such as Evans and Dunham that would cause them to rush toward a hand grenade or land mine and take the brunt of such an explosion to spare the lives of their fellow soldiers or Marines.

The only plausible explanation is they have a love for their fellow man that is so strong that it exceeds their natural instinct for self preservation. There have been many in the annals of the American fighting man, and there will be more. Thank God they will be there when our country needs them.

DESMOND "DEZ" BAILEY

"Let's check it one more time"

Dez Bailey in Army uniform

I first heard about Army Captain (now Lieutenant Colonel) Desmond "Dez" Bailey via my sister-in-law, Jerita Braun, a licensed nurse who often worked with Dez's mother Jenille in Montgomery. Jenille now runs a hospice service there.

Dez Bailey was born in Montgomery, Alabama. He and his brother Daxton were regulars at Morningview Baptist Church and enjoyed playing soccer as youngsters. His preference in music was the rock group "KISS," although his mom wasn't necessarily fond of that type music. Jenille had attended Troy State University 1967-70 and had been a member of the university's marching band under Dr. Johnny Long, a legend in his field. The music of Dez Bailey's generation was certainly a lot different from that of Tommy Dorsey and Glenn Miller enjoyed by veterans of World War II.

Dez attended private school at Trinity and Alabama Christian Academy in Montgomery until the tenth grade when he moved to Wetumpka to live with his father, Charles Bailey. Dez had a keen interest in sports and played cornerback and wide receiver on the Wetumpka *Indians* football team. The discipline and team effort lessons of football would help him later in his career.

In his senior year, much to his parents' chagrin, he decided to join the Marine Corps Reserves, so he completed basic training at Parris Island and incurred a 3-year commitment with the Marine Reserves in Montgomery.

He attended two years of college at Auburn University, Montgomery, then headed off to Troy State University where he subsequently made a decision to transition to the Army through Troy's then-new ROTC program.

Dez envisioned a career in the Army as an avenue to finance his schooling and have a meaningful life's work that would be beneficial to his fellow countrymen and women. He and four other cadets were the first to complete Troy's commissioning program in 1996.

His being in the Army was also a way for him to continue a long family tradition of military service. Jenille's father, Fred Vann, had been in the Army in WWII in the China-Burma-India theater for some four years where he fought with the *Gurkas*. Fred's brother Percy had served in the Navy and survived the Pearl Harbor attack. Fred's three other brothers—Gene, Grover, and William—had also served in the military in the World War II era. One of their uncles, Grover Cleveland Vann, had been an aviator in World War I. He had survived being shot down in France. The profession of arms runs deep in this family.

After his junior year of college, Dez married Kelly Hill of Holtville, and he and Kelly now have two sons, Jackson (age fourteen) and Cameron (age ten), as of this writing. The boys enjoy Scouting and martial arts training.

Graduating Troy, he went through training at Fort Benning's Infantry Officer—Basic, then completed Airborne school as well as Ranger training. When he made his first jump, he called his mom who asked Dez what it was like. He told Jenille that he shouted, "Thank you, Jesus!" when he landed safely on his first jump.

His first permanent assignment was at Fort Bragg with the 1st BN, 325th Airborne Infantry Regiment, where he began getting platoon leader time that included a 6-month rotation to Egypt from Jan-Jun 1998. Back at Bragg afterward, he served as a company XO and battalion S4 with the 325th.

He went back to Benning as a Basic Training Company CO (Bravo Company, 1 BN, 19th Infantry). As a senior lieutenant, he attended a Captain's Career Course, then left for Fort Hood to work in 4th Infantry Division Staff Plans. In 2001, he interviewed and got a job to lead Recon troops which would put him in a unique position to make history when he deployed to Iraq.

He moved to a job with 1st Brigade, 4th ID, where he was a planner for the entire brigade. In March, 2003, at the time Operation Iraqi Freedom was kicking off, he was the planner for their deployment.

He recalled that Turkey threw the USA a curve when they wouldn't allow troop movements through Turkey into Iraq, so the American forces had to be moved down to Kuwait to make their ingress. He recalled, "This required a lot of re-planning and shuffling of forces. A lot of the planning/re-planning had to be done on the hood of a Humvee in Kuwait." He took over as CO of a Brigade Recon Troop when they went into Iraq.

During this time in Iraq, he saw the palaces of Saddam Hussein and remembered how ornate they were until the US Air Force put JDAMs (smart bombs) into them. He was only able to call Kelly twice in the space of a year, and those calls were only for a minute or two. During his second call, he told Kelly, "Just watch the news."

By mid-December, 2003, Christmas was just around the corner, and his mom Jenille had her mind on preparing for the holidays. Cooked turkey and dressing, ham with candied yams, and all sorts of visions of sugarplums were dancing through her head. What came next was totally unexpected.

Commencing December 14, 2003, and for several days afterward, Jenille took calls from the local newspaper and television stations, her many friends, family and others who were simply curious about an interview her son had done with CNN from Iraq.

Dez was explaining a few things about how he and his G Troop, 10th Cavalry, 1st Brigade Reconnaissance Team soldiers helped track down Iraqi dictator Saddam Hussein and pull him from his "spider hole" after a search that lasted many months. I saw the replays later and wondered how they did it.

The entire country was relieved at the news when our Paul Bremer, head of the Coalition Provisional Authority, stood in front of news cameras, and, with a broad smile, announced, "Ladies and gentlemen, we got him," while breaking the news to an unsuspecting world.

Saddam Hussein was a ruthless tyrant who had once used weapons of mass destruction such as Sarin Gas, VX and other chemical weapons to inflict a horrifying death on 5,000 Kurdish men, women, and children in a small village in northeastern Iraq. His nuclear weapons aspirations had been blunted years before when Israeli aircraft took out his nuclear plant, but there was a fear he wanted to resurrect his program.

An international coalition of forces led by the United States military had invaded Iraq starting on March 20, 2003. On April 9th, American forces in Baghdad began pulling down statues of Saddam Hussein, signaling to the Iraqi population that the tyrant's regime was coming to an end. Saddam fled to parts unknown.

It was in the continuation of the Iraq War that then-Captain Dez Bailey became involved in what was termed *Operation Red Dawn*, one of the largest military manhunts in American military history that culminated in the capture of Saddam Hussein on December 13, 2003.

Members of a special task force that included elements of the 101st Airborne Division and Task Force 20 had killed Saddam's two sons, Uday and Qusay, plus a grandson, on July 22, 2003. In all, over 300 top leaders of the former government were killed or captured, as well as numerous lesser functionaries and military personnel.[1]

The task force assigned primary responsibility to find Saddam was led by Colonel Jim Hickey, CO of the 1st Brigade, 4th Infantry Division.

Reporting to Hickey was Lieutenant Colonel Steve Russell, who in June, 2003, took command of 1st Battalion, 22nd Infantry, and, in that capacity, commanded three companies and a HQ company, with close to 1,000 men in all. Ironically, Russell's assumption of command took place inside Iraq in Saddam Hussein's palace.

As CO of G Troop, 10th Cavalry, Dez Bailey led a Brigade Reconnaissance Team attached to Russell's outfit. As such, he had responsibility for scouting objective areas, setting up cordons, and responding to unforeseen contingencies.[2]

His team had, in his words, "...some great help from some of the finest Special Operations Forces in the world."

Saddam had been designated HVT#1—High Value Target Number One—and the success of the operation depended on capturing and interrogating numerous members of Saddam Hussein's family carrying the name Musslit, of whom there were ten brothers in a "Five Family Network." One of the Musslits was Saddam's personal bodyguard, and one had helped him escape Baghdad on April 9th, 2003. LTC Russell used a "Three Tier" strategy of working his way through Saddam's Musslit relatives to ultimately capture the top prize.

Each successive capture and interrogation of a Musslit family member yielded additional valuable intelligence. The hunt for Saddam was on, and Dez Bailey and his men were in the thick of it.

A great portion of Bailey's Area of Operation (AO) was along the Tigris River, both sides, near the towns of Tikrit (northwest of Baghdad), Ad Dawr, and Auja along Highway 1. On occasion, when attacked, they used .50 caliber machine guns and 40-mm MK-19 grenade launchers to repel the attacks. Bailey's AO consisted of a lot of farmland along the Tigris and was described by LTC Russell as, "So vast that it could only be patrolled, not fully covered. This lush farmland along the Tigris harbored rebels and resistance."[3]

A complex operation was set in motion on the night of Saddam's capture with conventional forces joined by Special Operations forces, with helicopters swooping back and forth above the action.

Bailey's troop set up a tight inner cordon around the farm where Saddam was suspected to be hiding. Realizing something was up, Saddam's cook and his brother spooked and fled into an orchard north of the suspected hideout. Bailey's troops apprehended the two, then joined SOF forces in scouring the areas based on scant info about an underground hiding facility.[4]

It appeared the search might be a dry hole, but Bailey didn't want to give up when other searchers thought it was fruitless. "Let's check it one more time," replied Dez.[5]

And they did, this time hitting the jackpot after a special operator treated rather harshly a Musslit that had been brought along as a source of intelligence. The Musslit pointed out where Saddam was hiding in his spider hole, and the long search was finally over, thanks in large part to the unceasing efforts of Army Captain Desmond "Dez" Bailey.

In March, 2004, Bailey came out of Iraq and took command of Charlie Company, 1st Battalion, 22d Infantry Regiment, for a year at Fort Hood. He was promoted to major in 2006. He then went to Fort Benning where he instructed Doctrine and Combined Arms Tactics in a Captain's Career Course for small groups.

An opportunity arose where he could become a battalion XO in a brigade in the 3rd ID at Benning, and he jumped at it. In Oct, 2006, he became an Armored Recon Squadron Ops Officer.

By March, 2007, he was back in Iraq for the "surge" that the Bush Administration advocated and was in the news so much. This tour in Iraq lasted 15 months and was tough, as his boys were growing up and he was missing out on many of the things they were doing. That type sacrifice is just another thing that career military men and women endure to serve their country.

Coming back, he had other jobs and assignments, to include Command and General Staff College as a student, as well as an assignment teaching at the School of Advanced Military Studies at Fort Leavenworth. Students at the school were required to read over 300 books and magazine articles to become steeped in the profession.

He was promoted to lieutenant colonel and went back to the 4th ID at Fort Carson, CO; the 4th was deployed to Iraq, and two weeks later Bailey was winging his way over to Kuwait for his third tour in Iraq. By late June, 2011, he took over as Division Chief of Plans with 4th ID HQ. He was in Iraq another five months, then came back to Fort Carson where he was still in that job until July, 2012, when he got a new job.

Lieutenant Colonel Bailey is presently commander of the 2d Battalion, 8th Infantry Regiment, in the 2d Brigade of the 4th ID. Their motto is, "First At Normandy," which is a rich legacy going back to WWII, and he has a very bright future with the Army, if he decides to stay a few more years.[6]

Not only is he a great leader, but he is a terrific teacher of young men and women. He stresses to his troops that, "You must learn lessons from the past so as to apply them to the future." He encourages them to read everything they can get their hands on regarding military operations, psychology, maneuvering, re-supply, planning for war, and many other facets of the profession of arms.

He says one thing that never changes is the tenant that, "You find the enemy first and shoot him before he shoots you." That's the first rule of survival, and he imparts that to his troops at every opportunity so that they too can survive the rigors of war.

Dez Bailey's mother, Jenille, his immediate family in the Alabama area, and others who knew Dez Bailey well during his youth are not surprised by his success.[7]

All patriots should be very proud of the role played by Alabamian Dez Bailey in tracking down and capturing one of the most despised tyrants in the history of the world. Thank goodness we have those of his determination on our side in this dangerous world.

PRISONERS OF WAR STORIES

"Man's most precious possession, second only to life, is freedom"

"I know what orders are, Major, and I am being hanged for obeying them," uttered Major Henry Wirz, Superintendent of the Confederate-run Andersonville Prison Camp (in Macon County, Georgia) near the end of the American Civil War.

Some 250 tickets were issued for spectators to observe the hanging of Major Wirz on November 10, 1865. When the trap door was sprung, the crowd chanted, "Wirz, remember Andersonville."

During its operation from February, 1864, to the end of the Civil War, some 45,000 Union prisoners of war passed through the Andersonville camp, with 12,913 dying of malnutrition, disease, or being shot trying to escape. That number represented 40 percent of all Union prisoner deaths.

Wirz was taken to Washington and put on trial for mistreatments of Union prisoners. He was the only individual tried, convicted and executed for war crimes in the entire Civil War. There easily could have been others.

The Confederates did not have a monopoly on running cruel prison camps. The Union's Camp Douglas facility housing Confederates near Chicago had some 6,000 deaths. John Thomas H. Lowe of Honoraville, Alabama, was one who survived Douglas, and his story is within this treatise.

The Union prison camp at Elmira, New York, was almost as bad. Camp Rathbun, dubbed "Hellmira" by inmates, in 15 months of operation held 12,100 Confederate prisoners. Of that number, 2,963 died of malnutrition, exposure to weather, and disease from unsanitary conditions.[1]

Throughout the history of warfare, prisoners have been mistreated. We have a lot of data and stories concerning American POWs and their treatment at the hands of Germans, Japanese, North Koreans and North Vietnamese. It wasn't pretty, by any means. Not only did they suffer physical and mental abuse, they often came home to uncertain futures.

American Arden Rowley was a POW who returned from the Korean War, having spent 2 years and 9 months as a prisoner. Other than the isolation and mistreatment, one of his worst memories was having to dig graves for fellow POWs who died in captivity. They were sometimes buried in graves hardly more than one-foot deep.

When the ship carrying Rowley passed under the Golden Gate bridge in September, 1953, his wife waiting on the dock was overcome with mixed emotions. On the one hand, she was fearful and didn't know what to expect. Yet, she was also joyous, and said, "It was the first time in my life that I cried because of happiness." Their story turned out well.

The Air Force's Bob Purcell didn't fare so well. He was held in North Vietnam 7 years and 6½ months. His first 17 months were in solitary confinement. He was imprisoned from July 27, 1965, to February 12, 1973.

One of the worst experiences for him was hearing fellow POWs being beaten and not being able to do anything about it. One thing he could do, though, was get food to a fellow prisoner, Jon Reynolds, who had been deliberately starved for eight days in 1966 for refusing the North Vietnamese guards' demands that he write an anti-war letter.

Purcell figured out a way to remove an enclosure in the ceiling of his cramped cell, crawl across the top, then remove the same enclosure in Reynolds' cell and drop a piece of bread to him, which saved his life.

Back home, his family thought Purcell was dead because his F-105 aircraft wing had broken off when he was hit over North Vietnam. They had a funeral for him and went on with life. Having him declared dead, his wife re-married.[2]

For POWs and families back home alike, missing important dates such as children's birthdays, anniversaries, weddings, and Christmas were the worst feelings. Those incarcerated and those who were free expressed this same thought.

Everett Alvarez was the first American aviator shot down and taken prisoner by North Vietnam. He was held the longest, 8 years and 6 months. His wife left him, which was devastating. The thoughts of his eventual family reunion were what kept him going and hopeful enough to survive the captivity. "My bubble burst, and it was just devastating" said Alvarez, upon hearing the bad news after his release.

Being deprived of food and nourishment caused dreams of food to dominate the thoughts of many in prison. One POW said he spent at least 10 hours a day planning in his mind the meals he would eat, if ever released. Another who loved Mexican food, said, upon his release, he ate Mexican for breakfast, Mexican for lunch, and Mexican for dinner, for one straight week.

Such a desire for good food can be appreciated, considering that in North Vietnam, prisoners were often fed weeds pulled from rivers and streams. Bread fed to prisoners in Germany had sawdust and leaves mixed in. Just enough sustenance to keep them alive. Some dropped to 120, even 100 pounds of body weight.

Sam Johnson, an Air Force fighter pilot and later a U.S. Congressman from the 3rd District in Texas, once made a speech where he said, "I spent nearly seven years as a Prisoner of War in Vietnam, more than half of that time in solitary confinement. I flew out of Hanoi on February 12, 1973 with other long-held Prisoners of War – weighing just 140 pounds... while in solitary confinement, my captors kept me in leg stocks, like the Pilgrims… for 72 days... As you can imagine, they had to carry me out of the stocks because I couldn't walk. The following day, they put me in leg irons, for 2.5 years. That's when you have a tight metal cuff around each ankle, with a foot-long bar connecting the legs."

Johnson continued, "I still have little feeling in my right arm and my right hand… and my body has never been the same since my nearly 2,500 days of captivity."[3]

David Eberly, the highest ranking prisoner of war taken in the Persian Gulf War, said his greatest angst was when he visualized a blue military sedan pulling up to his mother's home to tell her about his being captured.

William Seward, a prisoner of war of the Japanese in World War II, was trying to cope and figure out how to survive incarceration. A fellow POW told him, "Just keep your eye on the American flag."[4]

In his mind for the duration of his captivity, he visualized an America Flag. It was what kept him alive. One can imagine why a released POW such as Seward would be so repulsed at an anti-war protestor burning the American Flag, the image of which helped him survive his time as a POW.

Some prisoners were forced to dig grave-pits, then made to stand adjacent and face firing squads that, fortunately, used blanks in their rifles. The anxiety had to be overwhelming to endure such mental torture. Still, it was better than the alternative. And, they lived to fight another day.

To cope and maintain a sense of order and discipline, American prisoners in Vietnam used a "tap code," tapping on walls or windows or the floor to indicate a particular letter of the alphabet, and doing so in a manner to spell out certain words, phrases, or sentences. Or, in some cases, when allowed to sweep the sidewalk, some used a certain number of broom swishes to tap out the code and communicate with other POWs.

American Navy Vice Admiral James Stockdale was one POW who always looked forward to sweeping the sidewalk and carrying on a conversation with other prisoners in that manner. Communicating was crucial so inmates knew who was the highest ranking, which enabled them to maintain a chain of command, even under the worst of circumstances.

John McCain said later that giving up was the worst thing an inmate could do, and doing so greatly increased the chances of dying.

Some POWs described their ordeal as, "Hand-to-hand combat every day, but with your hands tied behind your back." Others said they would recall names, events, and dates of things they had not thought of in decades. Everything came flooding back in their isolation. One American POW learned French in his time in prison.

My friend Terri Newbauer's relative, Raymond Ison of Louisville, KY, was on the WWII Bataan Death March, then held POW in China three years. He slept on railroad tracks, ate only a little rice, and used his dog tags as a spoon. By the end, his weight had dropped to 100 pounds. He saw some POWs die on the day of their liberation. Ison was in the hospital for a year afterward.

Air Force fighter pilot Richard A. Dutton experienced the horrors of being a POW in North Vietnam. Dutton was born April 24, 1930, in Chicago, IL, to parents Geraldine and Ralph A. Dutton. He was commissioned as an Air Force second lieutenant through the Georgia Tech ROTC program in 1951.

In September, 1952, he graduated as a pilot from the Williams AFB, AZ program. After more training, he checked out in the F-86 *Sabre* and flew 59 combat missions in 1953 in the Korean War while assigned to the 336th Fighter Interceptor Squadron, which was part of the 4th Fighter Interceptor Wing.

After returning stateside from the war, he served as a gunnery instructor and checked out in the F-100 fighter aircraft. Later, during a tour in Germany with the 22nd TFS/36th TFW, 1960-63, he checked out in the F-105 *Thunderchief.* He later was an instructor at Nellis AFB, then deployed to Thailand where, with the 469th TFS at Korat, he flew 25 combat missions over Vietnam in the period March-June, 1966.

Along the way, he and wife Jean had a son, Russell, and a daughter, Diana, but Dutton would miss a large portion of their growing-up years. My wife, JoAnn, worked with Jean as a volunteer at the Hurlburt Field *Thrift Shop* in the late 1970s. I met Dutton once or twice at the Special Ops School.

Dutton came back to Nellis AFB, then was sent to *Wild Weasel* school prior to a second tour in Southeast Asia where he was flying with the 333d TFS out of Takhli Royal Thai Airbase from October 1967 to 5 November, 1967, when he was shot down.

He was captured in North Vietnam and would spend 1,956 days in captivity enduring extreme torture at the hands of his North Vietnamese captors. His Electronic Warfare Officer was Earl Glen Cobeil who was also captured and held as a POW. Cobeil was tortured to death by a Cuban interrogator referred to as "Fidel" by the Americans who experienced his extreme cruelty.

After his release, Dutton described what happened as follows, *"I landed right in the middle of a populated area. Quickly the peasants disrobed me with no thought of unfastening buttons or zippers. They even cut my boots. With elbows tied behind my back, a loose blindfold over my eyes, and a noose over my head, I was led barefooted down a rocky path. The civilians hit me with bamboo poles, rocks, dirt clods, and fists. I had a gaping wound, and one peasant woman stuffed it with a piece of cotton that had a mercurochrome-like antiseptic on it. Loaded into a small truck, we bounced along and finally arrived at an empty church. Shortly thereafter, communist soldiers put unconscious Glen Cobeil in one truck and me in another. Then we were taken to a Russian-built helicopter and placed in the cargo section. My ankles were tied to a floor hook. As we flew along my blindfold was pulled up around my forehead, and I saw an Oriental sitting on a packing crate holding a raised jack handle. I thought he was going to smash my brains in. He shoved my head around to look at Glen. There was no wound on him. We finally arrived at the Hanoi Hilton. Glen was alive. I never saw him again, but I heard him. We were tortured continuously, and, on the fifth day,*

I heard Glen scream my name and then I heard the sounds of them beating and clubbing him. He did not come home alive. They beat, they shackled, they even played a game of forcing me to sit on a stool in the middle of a room, and the guards would take turns with knocking me off the stool with blows to the head. The object was to see how far they could knock me. I think the record was ten feet! I have a loss of hearing in one ear due to those blows to the head. I existed from second to second, minute to minute, hour to hour. I would detach my mind from my body. I could take it and suspend a set of eyes above me and watch the show. I convinced myself it wasn't me being tortured. In fact, I wondered what they were going to do next to the poor soul. The torture went on seven nights and seven days. It was too much, I had to give in. However, we learned to bounce back and pass each hour in the hope that it would soon be time to go home."

Colonel Dick Dutton would not be released until March 14, 1973. After release, he spent time in a hospital to recover, then held assignments as Deputy Chief of Maintenance and later Commandant of the U.S. Air Force Special Operations School at Hurlburt Field, Florida.

Dutton retired September 30, 1981, after more than 30 years of service. He passed away on December 12, 1999.

Glen Cobeil's body was finally returned years after he had been declared dead on 5 Nov 1970. He is buried in Arlington Cemetery, and his place on the Vietnam Wall is Panel 29E, line 23.

"Fidel" the interrogator was felled by an assassin's bullet to the head years later in the parking lot of a Mexico City bar.[5]

Army Air Corps Captain Bert Bank's life changed drastically on April 9, 1942, when he surrendered to the Japanese in the Philippines and became a part of the *Bataan Death March*. Bank was born in Montgomery, completed high school in Tuscaloosa, and received a degree and Army commission from the University of Alabama ROTC program in 1937. As a death march prisoner, he and others were told they would not be allowed water or food. When he tried to get a drink from an artesian well, guards shot at him, but missed, and killed a Filipino standing next to him. He did get a sip from a mud hole at midnight the third day after a hard rain.

At noon each day, they were forced to reverse their march for some five miles, then turn and march forward again. Those who fell out were shot or bayoneted. Anyone with strength trying to help one who fell out was shot or bayoneted, along with the weaker one who fell by the wayside. In several cases, when prisoners passed out, the Japanese forced captives to dig holes and bury the weakened Filipinos and Americans, some of who were still alive. When Bank became weak and wobbly and bumped into a guard one night, he was struck on the head with an iron rod, rendering him unconscious for six hours. Unseen by guards, his fellow prisoners carried him through the night, and luckily were not shot or bayoneted themselves. He saw bodies decapitated alongside the march route. Some were burned instead of buried. In Camp O'Donnell where he was first held, some 26,000 Filipinos and 1,670 Americans died from starvation, beriberi, dengue fever, scurvy, dysentery, yellow jaundice, and other diseases. Over 60 per day died at the camp.

Those who wanted a drink of water had to stand in line for 10 hours to fill a canteen. Starving prisoners stayed close by those who were on the verge of dying so that they could grab the deceased's ration of rice at the moment of death. Most only wore clothing resembling a diaper. Some became slave labor in rice fields, working naked in water up to their necks, from 6 a.m. to 11 p.m., and developed sores all over their bodies. They were moved aboard "slave ships," and many died from suffocation when 12 prisoners were placed in a space designed for only two. Most lost weight down to 85-90 pounds and looked like skeletons.

He recalled that many gave up and didn't want to live. He was later, on two separate occasions, a POW at Cabanatuan where an average of 48 men a day died. Some 2,700 in all died there before 100 Sixth Army Ranger Battalion members raided the camp on January 30, 1945, and freed the 511 remaining prisoners.

Upon their release, for the fortunate, they described a general sense of elation. When Bert Bank and other released prisoners passed under the Golden Gate Bridge on their way home, he said there was not one freed prisoner aboard the SS *General Anderson* freedom ship that did not come on deck and cry at their joy to be back in the United States. Bank later in life secured the broadcast rights and built a radio network to air University of Alabama football games. He also was a state representative for two terms and a state senator for one term.[6]

Madeline Ullom, an American nurse held by the Japanese at Santo Tomas POW camp in the Philippines, remembered February 3, 1945, when American tanks broke through the front gates of the compound and a loud voice boomed "Hi folks! Are there any Americans in there?"

"Suddenly, the Camp was alive with voices calling out from everywhere. 'You'd better believe it!' 'We're in here!' 'We're Americans!'

"Some people were shouting 'They're here!' Others were yelling, 'We're free!' Someone started singing 'God Bless America' and, between laughing and crying, we all joined in..."[7]

In Europe, when POW camps were liberated, American GIs often found their guards had scurried away. British Army Captain Pat Reid was held prisoner at Colditz in Germany and described their liberation thus, "Suddenly, a mob was rushing toward him (an American soldier) shouting and cheering and struggling madly to reach him, to make sure that he was alive, to touch him, and from the touch to know again the miracle of living, to be men in their own right, freed from bondage, outcast no more, liberated, their faith in God's mercy justified, the nobility of mankind vindicated, justice at last accomplished and tyranny once more overcome. Men wept, unable to restrain themselves. Men with tears streaming down their faces kissed the GI on both cheeks—the salute of brothers."[8]

John McCain remembered upon his release he had a desire to read everything he could get his hands on. Reaching Clark Airbase in the Philippines, he grabbed every book, magazine, and periodical he could get his hands on and began to read vociferously, trying to make up for lost time.

Aircraft carrying American POWs home from Vietnam were described as "Freedom Birds," with those aboard describing scenes of pandemonium, singing, shouting, and jumping with delight. Scotty Morgan experienced elation he had not known. So did virtually all others aboard those C-141s.

American POWs depart North Vietnam on Freedom Bird (AF photo)

One prisoner released after the Persian Gulf War said the Americans provided two F-15 *Eagle* fighter aircraft that joined them inflight at the Kuwaiti border for escorting home. As the F-15s pulled up alongside the transport, each pilot did a victory roll as a show of respect and support. The POW said, "I knew we were 'home free,' then."

Retired Colonel Ben Purcell was the highest ranking Army POW during the Vietnam War. He was taken prisoner by the Viet Cong and spent 1,874 days as a POW in Laos—over 5 years. He escaped twice, but was recaptured both times. Much of the time was spent in solitary confinement, while enduring starvation and beatings.

Upon his release, Purcell stated, "Man's most precious possession, second only to life, is freedom." Purcell passed away on April 2, 2013.[9]

Another POW described his reunion with his children after they had grown so much as, "I felt like I was Rip Van Winkle," after having missed so many birthdays, anniversaries, sporting events, and other important occasions in their young lives.[10]

One POW that I was able to interview was Alabamian Roy McGinnis, now in Montgomery, Alabama. Born in 1923 to John and Myrtle (Sellers) McGinnis in Sylacauga, Alabama, McGinnis finished high school in Prattville in 1939. With four of his football buddies, they all joined the Alabama National Guard in 1940. He completed summer camp in Louisiana, plus drill on the week-ends. After one year, he volunteered for aircrew duty after watching a T-6 drop a flour sack target.

He re-enlisted for active duty, then completed aircraft mechanic school at Keesler Field, in Biloxi, after which he then worked mostly on

BT-13s. He later volunteered for and went to gunnery training at Tyndall in Panama City, ultimately becoming a gunnery instructor.

Assigned to an aircrew at a field near Ephrata, WA, he and his crew trained in Washington and Oregon. His pilot, Coles Goodner (of Tennessee), was the best pilot in their group, so, when assignments came down to fly a B-17 to Prestwick, Scotland, McGinnis was along with Coles' crew. They dropped the aircraft off at Prestwick then headed for Bovington, England, near London, where they arrived on September 7, 1943, for theater acclimation.

He was in the 96th Bomb Group, 339th Squadron, at Station 138, Snetterton. McGinnis was a waist gunner on the B-17 and was shot down on only his fourth raid. On his third sortie, a raid on Bremen, the Americans had 34 B-17s shot down. Immediately afterward, he wrote his mom and dad that he didn't see how he could make 25 missions and return safely.

On the next mission, on October 14, 1943, on what came to be known as *Black Thursday,* some 53 of the B-17s were shot down on the raids on Schweinfurt's ball bearing factories, with his among them.

This was the pivotal day when the decision was made that daylight bombing raids over Germany had to stop until better long-range escorts could be provided. American P-47s did not have range to go deep into Germany to escort the bombers all the way to the target and back. America simply could not produce enough bomber aircraft and train crews fast enough to continue on such a pace.

After losing an engine and falling out of formation, with Luftwaffe ME-110s firing on them, his pilot dove the plane toward the ground and made a controlled crash landing, sliding through mud and a pond to a stop, to avoid being completely shot out of the sky.

Flight Engineer Lester Kraft suffered a broken leg, and the mud-covered McGinnis and Navigator James Lapp hid in a corn patch until an elderly crippled man speaking Dutch beckoned them to come forward. They thought he was trying to assist them, but, instead, he called for the Germans who moved in and grabbed the crew as prisoners.

Young Roy McGinnis (his personal photo)

The officers and the enlisted were separated, as were Infantry and aviators, and all were taken to Frankfurt for interrogation. When all were lined up at the Frankfurt Railroad Station for further movement, it looked to McGinnis as if nearly the whole American Air Force had been shot down and captured.

Being young and naive, McGinnis thought he saw a young German girl smile at him, so he winked at her. She walked over, spat in his face, and the crowd of Germans nearly rioted, shaking their fists and yelling obscenities in German. Their guards had to lower their guns on the crowd to get them to calm down.

At the station, they saw rail cattle cars packed tightly with Jewish men, women and children passing through. Those who asked were told, "They're headed to *work camps*."

He noticed a difference in how prisoners were guarded. When later marching the Americans westward, German guards carried their weapons on their shoulders. When he and others passed them marching Jews eastward, the guards had their guns pointed at the Jews at all times.

A German who spoke perfect English interrogated him. He was told to fill out an International Red Cross "Kriegie" (prisoner) form. Trying to write Schweinfurt on the form, he spelled out "Swinefurd," and the German took it as an insult. The German demanded to know where Mc-

Ginnis's dog-tags were (in his rush to escape, he had left them in a flight jacket pocket on board the downed aircraft). He was quickly reminded that a "smart aleck" like him could be dragged off and summarily shot as a "spy," since he had no dog-tags. Some were hauled off to make an example for the others.

The aerial gunners were kept at *Stalag 17* near Krems an der Donau, Austria, 43 miles from Vienna, and were allowed one Red Cross package each week. The 12-pound packages had coffee, sardines, salmon, corn-beef, a block of cheese, crackers, a chocolate bar, and 5 packs of cigarettes, brand-names either *Camel* or *Lucky Strike*, most times.

As the war progressed and rail lines were bombed by the Americans, the packages arrived fewer and farther between. The German High Command began grabbing many of the packages for themselves. The prisoners' allotment of *Black Bread Broat* was one loaf per 3-4 prisoners. As conditions worsened, the allotment was one loaf per 25 prisoners (a daily slice). The bread contained 20% sawdust and 10% straw and leaves. Near the end, there were no bread loaves and few Red Cross packages.

Cigarettes were highly valued for either barter or gambling. They often were traded to guards for other items. Sometimes a pack of cigarettes could get one an egg or some other commodity.

The camp had a 50-hole latrine, and it filled fast. American aircrews were not required to work, but Russians were. Russians were made to empty the latrines, and the smell was quite odorous. POWs were allowed to bathe every three or four months and had to be de-loused periodically.

Two prisoners tried to escape through the snow by using undergarments over their clothing. The alarm sounded, and they were shot in the center of the compound, one killed and the other seriously wounded. After the escape attempt, guards came in the barracks and fired indiscriminately, punishing all for the "crimes" of two. One prisoner was shot in the buttocks on one such episode.

Some POWs literally went crazy, and many others screamed, yelled and reacted in other ways to the stress. Some had nightmares and cried out often in the night. One turret gunner was taken away.

McGinnis remembered one POW rushing to try to climb the fence with an American doctor trying to grab and hold him back, to no avail. The German guards cut him down with gunfire.

One constant was the cold. They were allotted 50 pounds of coal per week for heater warmth, but it never lasted an entire week. Sometimes, they were made to stand for hours in the snow. Winters were the worst, but in the Spring they could spend time outside and shake the bugs out of their straw-filled mattresses.

There were never enough blankets, and the POWs slept together to try to use body heat to stay as warm as possible. To counter the cold, when no guard was looking, they ripped outer boards off the barracks and burned them in the single small stove/heater in the center of the barracks. Almost all the exterior of the buildings disappeared by the time the guards began marching them westward for 18 days to stay ahead of the invading Russians approaching from the east.

Some skilled Americans traded items for radio crystals then constructed small radios where they were able to listen to the progress of the war via British Broadcasting Company (BBC) news programs. Progress of the war was always the one thing that generated interest among every POW.

McGinnis' group had a staff sergeant as commander, and he kept the others up-to-date on war progress, as the radio news was relayed to him.

The Germans were far more afraid of the Russians than the Americans, and, as the Russian approached, they marched the POWs westward for about 280 miles then had them hunker down in the woods. The Russians captured Vienna on 12 April, 1945.

He could hear Russian artillery explosions and see the flashes in the eastern sky as they approached. He saw pitiful sights such as Jews on marches who had not kept up and had been shot and left by the road.

At the end of the march, they were held in a wooded area near the Inns River where they tore the bark off trees to try to cover up. He first saw three tanks with American flags crossing. The tankers told them and their guards to stand pat and the American Infantry would catch up to them in 2-3 days for liberation.

When the Infantry arrived, they were told to go into Braunau, Austria—Adolf Hitler's birthplace—and take whatever food and shelter they needed. Once there, two American officers invited McGinnis and a buddy into their quarters they had occupied. Inside, he got his first bath in several months.

General Eisenhower had processing camps set up for POWs, and McGinnis was processed at *Camp Lucky Strike* in France where he was given food and new clothing. The POWs were liberated on May 5th and were back in New York on June 7th. McGinnis stayed sea sick almost all the way across the Atlantic Ocean. He was released in Philadelphia. His parents had moved, but the telegraph operator knew where to find them in their new location, so he forwarded the good news to them. After the war, McGinnis received $3,000 in back pay, left the service as a tech sergeant, got married, had a child, and spent the money.

In 1947, he came back in the Air Force as a staff sergeant and was at Maxwell AFB when Air University was formed. One of his First Sergeants at Maxwell was Ben Starr, father of Green Bay Packers quarterback Bart Starr. McGinnis went to Chanute, IL, to jet mechanic school, then back to Maxwell where he worked on transient aircraft, but was never really happy with the job. He was able to be re-classified into Personnel and went to typing school.

He went to Russian Language School at Syracuse University, and had other ROTC assignments at Cornell, St. Joseph in Philadelphia, and University of Pennsylvania. He pulled a tour in Thule, Greenland, too. McGinnis went to Phu Cat, South Vietnam, in 1966 for a year-long tour, then came back to Columbus AFB, MS, in 1967 and retired in 1968.

The next day he went to work for the State of Alabama, in the Claims Section of the Veterans Administration, and was there for 15 years until final retirement. Today, he lives in Montgomery, Alabama, with wife, Hilda.

McGinnis' crew on his B-17 were: Coles Goodner (Pilot), Arthur Leu (Co-pilot), James Lapp (Navigator), Howard Dey (Bombardier), Lest Kraft (Flight Engineer), Bernard Harman (Radioman), Roy McGinnis (Waist Gunner), John Cheney (Ball Turret Gunner), Melvin Jackson (Waist Gunner), Richard Ciurczak (Tail Gunner). [11]

Some POWs were not as fortunate as McGinnis and the others noted previously, such as nine who were held by the Japanese in Hiroshima. They perished in the firestorm when the first Atomic Bomb dropped in history from the B-29 *Enola Gay*, piloted by Paul Tibbets, Jr., killed 80,000 Japanese on August 6, 1945. They should not be forgotten.

At the Andersonville National Park is a plaque that was dedicated on June 27, 1985, with President Ronald Reagan's name inscribed. Those American POWs who died at Hiroshima were:

S/Sgt Charles O. Baumgartner USAAF
2LT Durden Looper USAAF
2LT James M. Ryan USAAF
Sgt Hugh H. Atkinson USAAF

Cpl John A. Long, Jr. USAAF
S/Sgt Buford J. Ellison USAAF
S/Sgt Ralph J. Neal USAAF
LTJG Raymond G. Porter USN
A/3C Normand Roland Brissette USN

The horror of being a POW is just another risk that has been faced by veterans of all conflicts. That risk will continue for all those who serve in future conflicts as well.

EPILOGUE

One of America's greatest baseball players once stood on home plate at Yankee Stadium and pronounced to the world that he believed himself to be, "The luckiest man on the face of the Earth."

This profound statement was made by Hall of Fame first baseman Lou Gehrig, who, at the time, was suffering from incurable Amyotrophic Lateral Sclerosis (ALS), a disease that was not only halting his stellar career, but, in short order, ending his life.

Gehrig felt privileged to have been born an American who was free to use his God-given talents to do whatever he wished with his life. All those born in America can and should make the same claim.

Their birthright is unsurpassed by the birthright of any other citizens of any other nations on the face of the Earth.

Most Americans recognize that to be true, but most especially those who have traveled abroad and observed the grinding poverty of numerous third-world nations, or who have witnessed or lived under the tyranny of governments ruled by dictatorial regimes. Perhaps we should ask "How did that come to be?"

I believe that birthright is so valuable because of our military men and women who served (and are currently serving) our nation so well over the centuries to ensure our values are preserved.

They were and are willing to "Pay any price, bear any burden, meet any hardship, support any friend, oppose any foe, in order to assure the survival and the success of liberty," as President Jack Kennedy so eloquently enunciated on a cold January day in 1961.

They will do whatever it takes to ensure the survival of our Republic, and we should always be ready to recognize their sacrifices, as well as the sacrifices of their families.

Their courage and willingness to valiantly fight and die for our freedoms and way of life is something worthy of preserving for posterity.

That's why I set out to record just a tiny fraction of the deeds of those in uniform.

Many of these heroes had been long-time acquaintances of mine, either socially or through my 26-plus years in the U.S. Air Force. Some were family members of friends who could provide information. Others became acquaintances through my contacting and interviewing them. The one thing we all share in common was that we wore a military service uniform.

My goal was not to just highlight those who were heroes by way of receiving a Medal of Honor or other high decoration. I hoped to record for posterity the service of many that I refer to as "ordinary citizens who performed extraordinary service" to their fellow man.

Whether they came off a farm, out of a textile mill, helped run a sawmill, had a paper route, or just ran errands for the corner drugstore, those who served in uniform made a lasting contribution for which all Americans should be extremely grateful. Some were poor and some came from privilege, yet their contributions were just as effective once they assumed the military responsibilities.

Most excelled not only in their military roles, but went on to serve society extremely well in post-military careers that often spanned decades. Some continued as military leaders, and some became educators, or politicians, or business leaders, or jewelry store owners, or mayors, or filled various and sundry other roles that made our country the great country that it is today.

Not all took the same risks, but it's possible they could have simply by virtue of the fact that they wore a military uniform. Because they contributed, they all should be recognized, and I hope this account has in some small way achieved that objective.

REFERENCES

Prologue

1. Article *Price For Our Freedom Was Paid For With Blood*, Columnist Cal Thomas, *Montgomery Advertiser*, Gannett Newspaper, July 5, 2013, pg 6A
2. Congressional Research Report-American War and Military Operations Casualties, updated Feb 26, 2010, and Annual Report, Navy and Marine Corps Military Personnel Statistics, June 30 1964.
3. *They Served Here*, Jeffrey C. Benton, AU Press, Maxwell AFB, July 1999, pg. 39-40

The Civil War

1. *The Eagle's Talons: The American Experience at War*, Col. Dennis M. Drew and Dr. Donald M. Snow, Air University Press, Maxwell Air Force Base, Alabama 36112-5532, Dec. 1988, pages 84-89
2. *Grant and Lee: A Study in Personality and Generalship*, Major General J.F.C. Fuller, Indiana University Press, Bloomington,©1957, pg. 19
3. *War,* Text by Albert R. Leventhal, Ridge Press, Inc., ©1973, pg 24
4. http://www.civilwarhome.com/casualties.htm accessed 11 May 2013
5. Ibid

Mathew M. "Mack" Rushing

1. *Perryville: This Grand Havoc of Battle*, Kenneth W. Noe, University of Kentucky Press, 2001, pp. 369, 373
2. *Grant and Lee: A Study in Personality and Generalship*, Major General J.F.C. Fuller, Indiana University Press, Bloomington, pg. 221
3. Source of some data: *Rushing Through the Generations*, A Genealogy by Peggy Rushing Sims of Glenwood, AL, 1994, pages 180-190

William Josiah "Diamond Joe" Ward

1. http://civilwartalk.com/threads/isaiah-boles-53rd-regiment-alabama-partisan-rangers.21966/ accessed Feb 2, 2013
2. http://www.littletownmart.com/family/Genealogy.asp?1049, accessed 2 Feb 2013 for Ward photo with wife

3. Source of story: Telephone Interview with Ralph Odell Ward, of Prattville, Alabama, 2 Feb 2013. Also, a prior visit to his home in Prattville, AL.

William Middleton E. Richburg

1. Info: Author's interviews with Charles Saunders and Dot (Scroggins) Saunders

Jonathan P. & John Thomas Henderson Low(e)

1. *Heroines of Dixie—Confederate Women Their Story Of The War,* Katherine M. Jones, Bobbs-Merrill, Co., Inc., Indianapolis/New York, 1955, pg 149-150
2. Source of much family information: Keith Lowery

World War I

1. *The 20th Century: An Illustrated History Of Our Lives And Times,* J.G. Press, Inc., © 2000, pg 127
2. *The Real War 1914-1918*, Captain B.H. Liddell Hart, Back Bay Books, Little Brown & Co., Boston New York Toronto London, ©1930, pg. 54
3. Ibid, pg 43

Willie Lee Rushing

1.http://www.encyclopediaofalabama.org/face/Article.jsp?id=h-1545 accessed 11 May 2013
2. *Rushing Through the Generations*, A Genealogy by Peggy Rushing Sims, 1994, pg 207

Marion Washington Lowery

1. Alabama State Archives, accessed 26 Feb 2013
2. Ibid
3. Family information from Keith Lowery

Lemon O. Lee

1. Alabama State Archives, accessed 6 Mar 2013

Robert Frank Lawrence

1. Alabama State Archives, accessed 8 Mar 2013
2. Much family information from: Don Newbauer, Jr., Cincinnati, Ohio

World War II

1. http://www.pearlharbor.org/speech-fdr-infamy-1941.asp, accessed Jan 16, 2013
2. http://www.ww2f.com/topic/14702-dorothy-edgers-and-pearl-harbor/ accessed 20 May 2013

Charles Saunders

1. *The 20th Century: An Illustrated History of Our Lives and Times,* article: "Judging Hitler's Henchmen," JG Press, North Dighton, MA, ©2000, pg. 347
2. Author's interview with T.J. Boothe on April 12, 2013, at his home, Opp, AL
3. http://en.wikipedia.org/wiki/Atomic_veteran acc 12 Feb 2013
4. Author's multiple interviews with Charles Saunders, Troy, AL, as well as a telephone interview with his sister-in-law, Dorothy (Scroggins) Saunders

Richard E. Cole

1. *Destination: Tokyo A Pictorial History of Doolittle's Tokyo Raid*, Stan Cohen, Pictorial Histories Publishing Co. Missoula, Montana, 13th Printing, March 2012, pg 36
2. *Winged Victory: The Army Air Forces in World War II,* Geoffrey Perrett, Random House, New York, 1993, pg 151
3. *Flyboys: A True Story of Courage*, James Bradley, by Little, Brown & Co., N.Y., N.Y., page 105
4. Author interview with Lt Col (Ret) Richard E. Cole, Ft. Walton Beach, FL, Oct, 2012
5. http://en.wikipedia.org/wiki/Jimmy_Doolittle, accessed 4 Feb 2013
6. Interview Cindy (Cole) Chal on 17 April, 2013, Eglin AFB Armament Museum

James Ralph "Shug" Jordan

1. http://www.encyclopediaofalabama.org/face/Article.jsp?id=h-1502
2. Alvin Benn, guest columnist, *Montgomery Advertiser*, June 6, 2010
3. http://www.archives.state.al.us/famous/academy/r_jordan.html
4. Benn, *op cit*
5. http://www.archives.state.al.us, *op cit*

Tuppie Leroy Rushing

1. Source of some data: *Rushing Through the Generations*, A Genealogy by Peggy Rushing Sims, 1994, pages 57-60.
2. Newspaper account of Florence Times accessed on line 29 Jan 2013 at: http://news.google.com/newspapers?nid=1842&dat=19450206&id=J_srAAAAIBAJ&sjid=I-LoEAAAAIBAJ&pg=3741,1490715
3. Photo courtesy of Margie (Rushing) Williams, Ponce de Leon, FL
4. Information from other stories told by Tuppie were courtesy of Tuppie's son Eddie Lee Rushing and daughter Anita (Rushing) Mitchell

Arthur Robert Scroggins

1. http://www.usafa.edu/uploads/temp/zemkerpage.cfm, accessed 6 July 2013
2. http://en.wikipedia.org/wiki/Cologne, accessed 12 Feb 2013
3. Author's telephone interview 12 Feb 2013, plus visit to Scroggins' home for second interview on 16 Feb 2013

Ralph Odell Ward

1. http://www.uscg.mil/history/webcutters/img/LST207_1_300.jpg
2. http://www.bing.com/search?q=LST-207&FORM=O1HV1, accessed 29 Apr 13
3. Various telephone calls and a visit by author to Ralph Ward's home 1 Feb, 2013

Charles Durning (Actor)

1. http://www.nndb.com/people/939/000022873/ accessed Feb 5, 2013
2. Source of information: Article by AP writer Bob Thomas, Dec 25, 2012

Ralph, John, and Tony Peacock

1. http://en.wikipedia.org/wiki/RAF_Great_Ashfield accessed 5 May 2013
2. *Winged Victory: The Army Air Forces in World War II*, Geoffrey Perret, Random House, New York, 1993, pg. 365
3. Grady County, GA, Memorial website, accessed 22 Mar 13
4. Interview with Tony Peacock in Milton, FL on January 12, 2013

Charles P. "Chuck" Hundley

1. http://www.nationalmuseum.af.mil/factsheets/ factsheet_print.asp?fsID=1664&page=1 accessed 7 Feb 2013
2. *Over The Hump,* William H. Tunner, Duell, Sloan and Pearce, ©1964, and reprint by U.S. Government Printing Office for U.S. Air Force studies, 1985, page 9
3. Ibid, pg 185
4. http://en.wikipedia.org/wiki/Berlin_Blockade accessed 7 Feb 2013

William Nelson Murdock

1. Recollections of Rebecca Jean (Adams) Murdock, received 9 May 13
2. Ibid
3. Ibid
4. email and phone interview with William Nelson Murdock, Jr., 9 and 10 May 13

Robert T. Bennett

1. http://acepilots.com/planes/p51d_mustang.html, accessed 6 Jul 2013
2. http://en.wikipedia.org/wiki/B-47_Stratojet#Specifications_.28B-47E.29
3. Author interview with Robert T. Bennett, Jan 22, 2013, in Georgiana, AL

Hubert T. Sullivan

1. *Citizen Soldiers*, Stephen E. Ambrose, Touchstone (Simon & Schuster), Rockefeller Center, New York, N.Y., © 1997, pg. 408
2. Author's personal interview with H. Sullivan at his home on Jan 20, 2013

James Forrest "Frosty" Cook

1. http://en.wikipedia.org/wiki/USS_New_Mexico_(BB-40)
2. Author's interview with Frosty Cook January 8, 2013, at his home in Opp, AL

Morgan M. Moseley

1. http://en.wikipedia.org/wiki/USS_Indianapolis_(CA-35)
2. Interview with author Oct 7, 2012, at Moseley's home

Wilson Merrill Sauls

1. *Citizen Soldiers,* Stephen E. Ambrose, Touchstone (Simon and Schuster), Rockefeller Center, New York, N.Y., © 1997, pages 433-434

2. *A Paratrooper's Panoramic View,* Robert L. Wilson & Philip K. Wilson, AuthorHouse, First Edition, edition (Dec 5, 2005)
3. 17th Abn Div History, by Bart Hagerman, accessed on internet 11 Feb 2013
4. Author's interview with Wilson Merrill Sauls at his home January 19, 2013

John Cluck, Jr.

1. *The War in the Air: A Pictorial History of World War II Air Forces in Combat,* Gene Gurney, Major, USAF, Bonanza Books, division of Crown Publishers, Inc., 419 Park Avenue S., New York, N.Y., pg 320
2. *The Pacific War: The Strategy, Politics, and Players That Won The War,* Hopkins, William B. (2009), Zenith Imprint, ISBN: 0-7603-3435-8, pg 325
3. Interview with author at VFW Post 96 meeting, Montgomery, AL 28 Jan 2013. A second interview was conducted at his home on February 10, 2013

Hunter "Bill" Maxwell Bishop, Jr.

1. http://www.gradymilitary.us/detail-viewview.php?id=217 accessed 5 Feb 2013
2. 505th Parachute Regiment Unit History
3. July 11, 1980 Cairo *Messenger*
4. http://en.wikipedia.org/wiki/Sainte- M%C3%A8re-%C3%89glise
5. http://www.gradymilitary.us/detail-viewview.php?id=217 accessed 5 Feb 2013

John Patrick "Jack" O'Brien

1. http://en.wikipedia.org/wiki/27th_Troop_Carrier_Squadron, accessed 11 Feb 2013
2. Telephone interviews with Jack O'Brien 9 and 23 Feb 2013, and information mailed to author by O'Brien

Evon Williamson

1. Author interview with Williamson and family, Gantt, AL, May 8, 2013

Lee Jack Lowery

1. Interview with and various emails from Keith Lowery

Noah Nelson Cowen

1.http://airforce.togetherweserved.com/usaf/servlet/tws.webapp.WebApp?cmd=ShadowBoxProfile&type=Person&ID=86794
2. Cowen relatives Janice Lambert and Joan Reynolds provided info and two photos.

Thomas Dwight Teel

1. *The Opp News*, Thursday, August 17, 1944, pg. 1
2. *The Opp News*, Thursday, September 14, 1944, Covington County Boys in Service
3.http://www.littlefriends.co.uk/356thfg.php?sort_order=ASC&order_by=Pilot&recs=50&next_page=1

4. Charles Saunders' recollections regarding Teel, as well as recollections of Teel's brother, Thomas H. Teel

Capt. Writes Opp Mother of Son's Death (Grady Colvin)
1. *The Opp News*, Thursday, Sept. 14, 1944 copy provided by Janice Lambert
Korean War
1. *American Caesar—Douglas MacArthur 1880-1964*, by William Manchester, Dell Publishing, 1540 Broadway, N.Y., N.Y., copyright 1978, pg. 663
2. Tucker, Spencer C., ed. *Encyclopedia of the Korean War: Political, Social, and Military History*. Santa Barbara, CA: ABC-CLIO, 2000
3. http://en.wikipedia.org/wiki/Korean_War_casualties#Casualties acc 28 May 13
4. *From Lexington to Desert Storm: War and Politics in the American Experience,* Donald M. Snow and Dennis M. Drew, M.E. Sharpe, Inc., New York, ©1994, page 180

Millard Martin
1. Author met and talked with Mr. Martin at Maxwell AFB, Alabama, September, 2012. Also, a telephone interview on Jan. 23, 2013 and a second phone call on 25 Feb 2013
Kurt Miller
1. Author's Interview with Miller on 28 Jan 2013, and subsequent phone call
Harold O. Chitwood
1. *Aviation History* magazine, Diego F. Zampini, Leesburg, VA, Nov. 2005, pg 35
2. http://www.check-six.com/Crash_Sites/F-86H_crash_site.htm
3. Author interviews with Harold Chitwood on 24 March and 6 May, 2013

Roy H. Johnston
1, *A War Too Long: The USAF in Southeast Asia 1961-1975,* John Schlight, Air Force History and Museums Program, 1996, pg. 42
2. Author personal interview with Johnston

Joel E. Lawrence
1. Interview with and photos/news clippings from Don Newbauer, Jr.
Lt Gen Charles G. "Chick" Cleveland
1. *Once A Fighter Pilot: The Story of Korean War Ace Lt .Gen. Charles G. "Chick" Cleveland,* Warren A. Trest, River City Publishing, Montgomery AL, ©2012, pg 60
2. Author's Interview with General Cleveland 12 Apr 13

Jim, Art, Don, Charles, and Clarence Newbauer
1. Jim Newbauer 24 Nov 1950 letter to parents—courtesy of Don Newbauer, Jr.
2. Contents of newspaper article obituary courtesy of Don Newbauer, Jr.
3. Photo courtesy of Don Newbauer, Jr.
4. Source of other family info: Don Newbauer, Jr., Cincinnati, OH

William "Bill" Dudley, Carlton, and Bob Taylor

1. Author's telephone interviews with Maureen Taylor, May 16 and 17, 2013

2. Author's telephone interview with Bob Taylor, Opp, AL, on May 7, 2013

Douglas Campbell

1. Douglas Campbell memoirs, made available to author May 12, 2013
2. Ibid
3. Ibid
4. Ibid

Luther H. Story

1.http://projects.militarytimes.com/citations-medals- awards/recipient.php?recipientid=2282 (source of photo from Hall of Valor site)
2. http://en.wikipedia.org/wiki/Luther_H._Story
3. http://www.findagrave.com/cgi-bin/fg.cgi?page=gr&Grid =7745889

Vietnam War

1. *From Lexington to Desert Storm: War and Politics in the American Experience,* Donald M. Snow and Dennis M. Drew, M.E. Sharpe, Inc., New York, ©1994, pages 215- 217

Sir Francis McGowin

1. Author interview with Sir Francis McGowin in his home Andalusia, AL 8 Jan 13, and information from his son Steve on Jun 27,2013

Durwood Lamar Gooden

1. Author interview with Gooden conducted January 19, 2013

Gene Connor

1. Author interview with Connor on 28 Jan 2013 at VFW Post 96 meeting in Montgomery, AL, and subsequent visit to Gene's home for second interview

Robert McCarty

1. Author interview with McCarty at his home

David J. Myers

1. Author interview with David Myers

Rodney John Evans

1. Information from David E. Robertson, boyhood friend of Evans
2. http://en.wikipedia.org/wiki/Rodney_J._Evans
3. Ibid

4. Ibid

Jim and Patricia Pochurek

1. email from Jim Pochurek 3 Mar 13
2. ACA Quarterly Membership Newsletter, Dec, 2012, pg. 10
3. *In The Midst Of Wars,* Major General Edward Geary Lansdale USAF (Ret), Harper & Row, New York, ©1972, pg 50-51
4. Ibid, pg 151, 4th paragraph, under Notes
5. Visit with Pochureks in Ft. Walton Beach, FL, Oct 2012

Charles "Chuck" Manikas

1. http://texashistory.unt.edu/ark:/67531/metapth254878/m1/8/zoom/
2. Interview with Chuck Manikas at his home in Montgomery, AL

James and Barbara Jarrett

`1. Interview with Jim and Barbara Jarrett at their home in Montgomery, AL

Donald Jack Neese

1. DFC Citation, Military Hall of Valor website, accessed 23 Mar 2013
2. Silver Star Citation, Military Hall of Valor website, accessed 23 Mar 2013
3. Donald J. Neese biography, courtesy of Opp (AL) Chamber of Commerce, and Neese telephone interview with author on 25 March 2013

Herschel Scott Morgan

1. http://www.pownetwork.org/bios/m/m119.htm
2. http://www.pownetwork.org/bios/a/a044.htm

Post-Vietnam/Pre-Desert Storm

1. *Operation Thursday: Birth of the Air Commandos*, Herbert A. Mason, Jr., SSgt Randy G. Bergeron, TSgt James A. Renfrow, Jr., Air Force History and Museums Program, 1994, pg 25

Major (Ret) Logan Fitch

1. Logan Fitch's email to author on 26 May, 2012
2. http://www.history.com/this-day-in-history/173rd-airborne-brigade-deploys-to-south-vietnam
3. *Airman's Odyssey: An Air Force Special Operator's Incredible Journey*, Lt Col James D. Lawrence, USAF (Ret), Deeds Publishing, Marietta, Georgia, ©2012, pages 179-180
4. email from Logan Fitch to author May, 2012

Bert Langford

1. *Air Force Magazine*, Mar, 99, article El Dorado Canyon, by Walter J. Boyne, pg. 58
2. http://en.wikipedia.org/wiki/Gulf_of_Sidra_incident_(1981)
3. Author's interview with Bert Langford, April 2, 2013, Sandestin, FL

James Jones

1. Author interview with Jones in Onycha, AL, April, 2013

Maj Gen (Ret) James L. "Jim" Hobson, Jr.

1. http://www.af.mil/information/bios/bio.asp?bioID=5814
2. http://militarytimes.com/citations-medals-awards/recipents.php?
3. http://www.af.mil/information/bios/bio.asp?bioID=5814/ *op cit*

Persian Gulf Wars

1. http://en.wikipedia.org/wiki/Invasion_of_Kuwait accessed 18 May 2013
2. *From Lexington to Desert Storm: War and Politics in the American Experience, op cit,* pages 296-297
3. Ibid

Michael Patrick and Cheryl L. Hynes

1. emails from Michael Hynes May 20, 2013, and personal knowledge

Patrick Brian Guris

1. http://en.wikipedia.org/wiki/Aviation_ Machinist%27s_Mate*, accessed 8 Apr 13
2.http://en.wikipedia.org/wiki/USS_Saratoga_(CV-60), accessed 9 Apr 13
3.http://www.af.mil/news/story.asp?id =123264068, accessed 8 Apr 13
4.http://www.nwfdailynews.com/military/eglin/f-35-production-on-track-1.114994, accessed 8 Apr 13
5.Telephone Interview with Pat Guris April 21, 2013
6. Interview with Jim and Barb Guris, Mary Esther, FL, April 13, 2013

Bob and Guy Buice

1. Author's telephone interview with Bob Buice, Jan, 2013

Corporal Jason L. Dunham

1. *The Gift of Valor: A War Story,* Michael M. Phillips, Broadway Books, New York, pages 99-108
2. Ibid, pg. 134
3. Ibid, pg. 226-227
4. Ibid, pg 227
5. http://en.wikipedia.org/wiki/Jason_Dunham, accessed March 8, 2013

Desmond "Dez" Bailey

1. Pentagon: Saddam is POW, CNN, Jan 10, 2004
2. *We Got Him!: A Memoir of the Hunt and Capture of Saddam Hussein*, Lt. Col. Steve Russell, U.S. Army (Ret), Deeds Publishing,©2011, pg. 90
3. Ibid, pg. 223
4. Ibid, pg 319
5. Ibid, pg 320
6. Author telephone interview with Dez Bailey May 24, 2013

INDEX

7. Various info from author's interview with Bailey's mother Jenille, 30 Apr 13

Prisoners of War—Their Stories

1. http://web.archive.org/web/20081202041228/http://www.cityofelmira.net/ history/prison_camp.html accessed Jul 8, 2013
2. http://www.greasyonline.com/article203.html
3. http://samjohnson.house.gov/news/documentsingle.aspx?DocumentID=58470
4. Video display at Andersonville, GA, Civil War Prison National Park
5. Information on Richard A. Dutton was compiled by W.H. Plunkett, Albuquerque, NM, printed 31 May, 2013. Plunkett did extensive research of Air Force records, as well as interviews with other knowledgeable individuals.
6. *All Of Us Fought The War, The University of Alabama and its Men and Women in World War II*, by Delbert Reed, published by Paul W. Bryant Museum, Tuscaloosa, AL, ©2012, pages 57-67 (Bert Bank wrote the book *Back From The Living Dead* in 1945, and Reed used some information from that book about Bank's POW years).
7. *All This Hell: U.S. Nurses Imprisoned by the Japanese*, by Evelyn M. Monohan and Rosemary Neidel-Greenlee, University of Kentucky Press, ©2000, pg 155
8. *Citizen Soldiers*, Stephen E. Ambrose, Touchstone (Simon & Schuster), Rockefeller Center, New York, N.Y., © 1997, pg. 459 (quoting from British Captain Pat Reid's book *The Latter Days at Colditz*, pg 226)
9. *Rolling Thunder Washington, D.C. Inc*
10. Author's Note: Several POW quotes were from a visit to the Andersonville, GA, POW museum which has recordings. (Those of Everett Alvarez, Arden Rowley and wife, David Eberly, John McCain and others are among this group).
11. Author interview with Roy McGinnis, Montgomery, AL, March 19, 2013

ABOUT THE AUTHOR

Retired Air Force Lieutenant Colonel James D. "Jim" Lawrence, is a native of Opp, Alabama, OHS Class of 1965, and a graduate of Troy University (B.S. in Accounting/Bus. Administration, and Master's in Public Administration).

He was a Command Pilot with 5,800 hours flying time over a 26½—year career. He flew as an instructor and check pilot in the T-38 *Talon* in Oklahoma, instructor in airlift C-130s in Okinawa and Japan, and was an instructor and flight examiner in both the AC-130H *Spectre* Gunship at Hurlburt Field, Florida, and the HC-130P/N in Aerospace Rescue and Recovery at Eglin Air Force Base.

For many years, he was co-holder of the world record for C-130 non-stop flight time (29.7 hours, set in November, 1979). Eighteen of his active duty years were in Special Operations, to include not only flying, but assignment as a charter member of a joint command staff in a highly-classified organization.

In 1980, he participated in the attempted rescue of American hostages held in Iran, and later flew a support mission in an effort to free AP journalist Terry Anderson who was held in Beirut. He searched for and located pieces of the exploded Space Shuttle *Challenger*. In his career, he flew 23 different military and civilian aircraft, traveling to more than 40 countries and islands.

Eight of his years were in the academic world, where he was an instructor of Unconventional Warfare and Crisis Response Management courses, then Operations Division Chief, Vice Commandant, and interim Commandant of the U.S. Air Force Special Operations School, Hurlburt Field.

In a post-Air Force career, he worked more than 16 years as an inspector and consultant in the Environmental Field. He and wife JoAnn have two children and three grandchildren, and they currently reside near the Montgomery, Alabama area.

He previously authored the book: *Airman's Odyssey: An Air Force Special Operator's Incredible Journey, ©August, 2011,* Deeds Publishing, Marietta, GA.

www.ingramcontent.com/pod-product-compliance
Lightning Source LLC
LaVergne TN
LVHW052036200926
848481LV00005B/1695
* 9 7 8 1 9 3 7 5 6 5 7 3 2 *